U0897090

Small Sample Houses

小户型样板房

深圳市智美精品文化传播有限公司 主编

華中科技大學出版社
http://www.hustp.com
中国 · 武汉

图书在版编目（CIP）数据

小户型样板房 / 深圳市智美精品文化传播有限公司 主编 . – 武汉 : 华中科技大学出版社 , 2012.11

ISBN 978-7-5609-8191-8

Ⅰ . ①小… Ⅱ . ①深… Ⅲ . ①住宅 – 室内装修 – 建筑设计 Ⅳ . ① TU767

中国版本图书馆 CIP 数据核字（2012）第 153471 号

小户型样板房　深圳市智美精品文化传播有限公司 主编

出版发行：华中科技大学出版社（中国 · 武汉）
地　　址：武汉市武昌珞喻路1037号（邮编：430074）
出 版 人：阮海洪

责任编辑：段自强　责任监印：秦英
责任校对：段园园　装帧设计：百彤文化

印　　刷：利丰雅高印刷（深圳）有限公司
开　　本：992 mm × 1240 mm　1/16
印　　张：20
字　　数：160千字
版　　次：2012年11月第1版 第1次印刷
定　　价：298.00元（USD 59.99）

投稿热线：（020）66638820　1275336759@qq.com
本书若有印装质量问题，请向出版社营销中心调换
全国免费服务热线：400-6679-118 竭诚为您服务

WONDERFUL DUE TO SMALLNESS
小得精彩

The dwelling narrowness egg house, capsules and other agnames extremely exaggerate the delicate characteristics of small size of house. In fact, the people's needs are really less, only including a table, a cabinet, a bed and a room. Consumption is high, price of house is expensive, and thereby pressure is intense, people has no place to go at the flourishing city, run away from Beijing, Shanghai and Guangzhou become an negative effect of flourishing city, Fortunately, there can be a life named small size of house ranging from 15 square to more than 30 or 40 square capable of accommodating a table, chairs, a bed, cabinets, kitchen, bathroom, learning and living, or even work, it is full of hope, you do not need to pay high rent, hot meals wait for you after work. It is an occupation step for tired recreation as well as cornerstone of fighting.

In contrast, Hong Kong, Singapore and Japan, place without sufficient land resources, are accustomed. Recently, a set Japan design program "Super Transformation King" records a whole process of transforming small three-floor house with the area of 60 square meters accumulating five people into 80 square meters. The people think that they can upgrade the house from 60 square meters to 80 square meters, and every one can get what he wants, and they are extremely happy. People in Hong Kong believe that houses of more than 60 square meters are common, even Zhang Zhiqiang, a famous architect, is keeping his "small house" with 60 square meters, and he also designed it abnormally vivid, the sitting room, study, kitchen and cabinets are can completed contained in the row cabinets which can be combined with the aid of the file cabinet theory. The small size of house in our eyes is actually the most common appropriate room for people in Hongkong, Singapore is located in the tropics, small size of house has open spaces except kitchen and toilet, and other rooms are connected with the big balcony, ventilation requirement is very high with concise shape and elegant style, and all functional spaces have high degree of flexibility.

In prosperous cities, small size of house is placebo for older generations as well as doping for younger generation. It is a small boat, but it can have own strong power and firm direction.

Chen Zhibin

蜗居、蛋屋、胶囊等等绰号极尽夸张地把小户型的玲珑特点呈现出来。其实，人的需求真的不大，一桌、一柜、一床、一间房而已。消费之高，房价之贵，压力之大，在繁华都市似乎已无可容身之所，逃离北上广成了繁华的负效应。幸而能有一种生活叫小户型，少则15平方，多也不过三、四十平方，容得下桌、椅、床、柜、厨、卫、学习、生活，甚至工作，充满着希望，不用付高房租，下班有热饭菜。劳累之憩所是职业之台阶、拼搏之基石。

相比之下，香港、新加坡、日本这些天生缺少土地资源的地方则习以为常。最近一集日本设计类节目《超级改造王》里面记录了一个住了五口人的三层约60平方的小住宅改造成80平方的全过程。这家人认为能从60平方升级到80平方的面积，且一家人能够从老到小各得其所都高兴极了。而在香港朋友的眼里，60多平方的住宅遍地都是，连著名建筑师张智强都一直保留着自己的60平方的“小家”，还设计得异常生动，借助档案柜的原理，把客厅、书房、厨、柜全盘收纳到这个可合并的排柜里面。我们眼中的小户型，其实是港人最常见的经适房啊！新加坡地处热带，小户型里除了厨、卫，其余都敞开，并且与大阳台相连，通风要求甚高，造型简洁，格调淡雅，各功能空间具有高度的灵活性。

在繁华的都市中，小户型是年老一代的安慰剂，更是年青一代的兴奋剂。它是一叶小舟，却能有自己不挠的动力和坚定的方向。

陈志斌

CONTENTS / 目录

紧凑小复式
THE COMPACT DUPLEX

精致小户型
REFINED SMALL DWELLING-SIZE APARTMENTS

迷你型公寓
THE MININATURE APPARTMENT

006–101
COLOMBO
STAR WARS

THE COMPACT DUPLEX

紧凑小复式

设计公司：深圳市昊泽空间设计有限公司 | 设计师：韩松 | 项目地点：中国江苏 | 项目面积：50 平方米 | 主要材料：木地板、石材、银镜

HUAYANGNIAN REAL ESTATE WUXI XINIAN CENTER LOFT.A TYPE

花样年地产无锡喜年中心 LOFT.A 户型

A home is desired as a harbor of the soul, and a destination to own soul. This is a space which is not too big without gorgeousness, this is a place which can provide new beginning, wherei n one can easily work and pleasantly live. It can give you a cup of hot coffee when you are listless, and can afford to dancing when you are happy, it only belongs to you and her (his).

The designer takes it as concept to show his design enthusiasm in the space of 50 square meters, thereby melting simple, stylish and artistic temperaments into the space, people can enjoy their freedom and happiness. Ceilinged advantage is utilized to produce the interlayer, cantilevered stair treads are matched with it, additionally, the interlayer forms semi - surrounded structure with transparent glass to form a light and elegant feeling. Interior color is mainly based on white and wood color dotted with black frame and personalized lighting so that the space is filled with a sense of art. It is wonderful to live here.

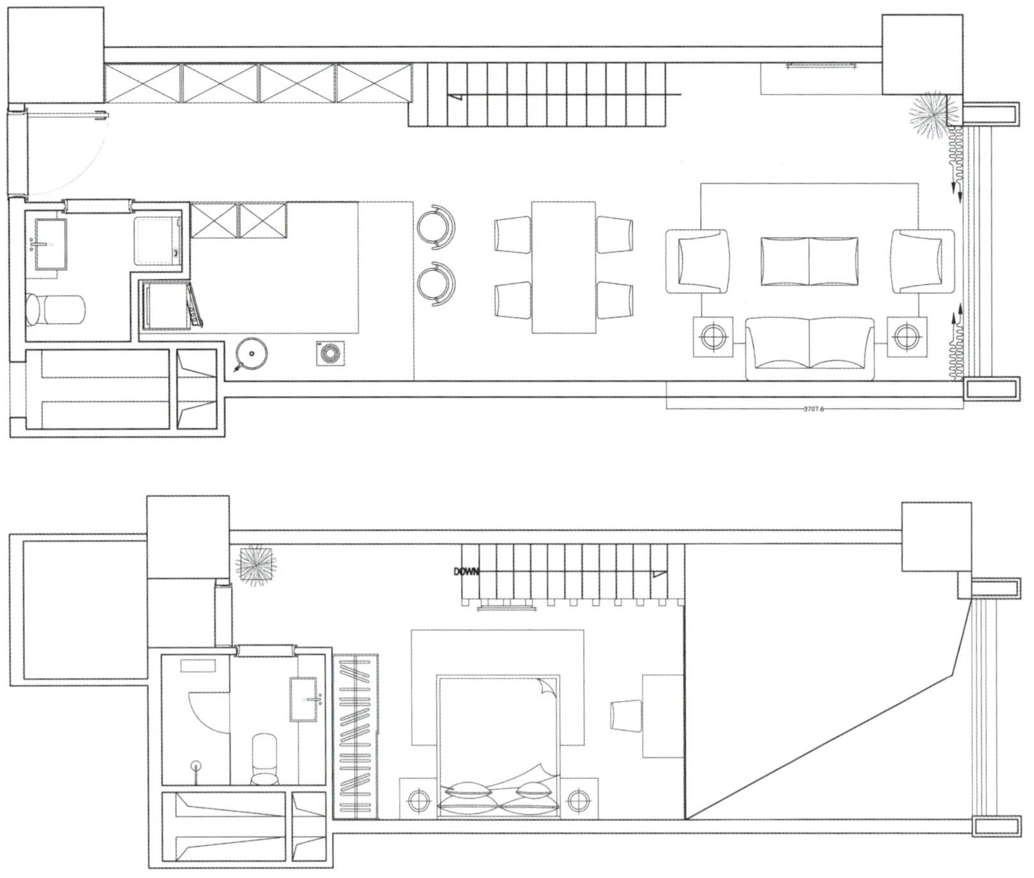
DOWN

想有个家，作为心灵的港湾，给自己的灵魂找一个归属地。这是一个不需要太大、无需华丽的空间，这是一个提供全新的开始，一个可以轻松工作、愉快生活的地方。在你无精打采的时候可以给你提供一杯热咖啡，在你高兴时可以毫无顾忌地手舞足蹈，一个只属于你和她（他）的地方。

设计师以此为念，在这个 50 平方米的空间里，尽情挥洒设计的热情，将简约、时尚、艺术的气质融入空间中，让人尽情享受属于自己的自由和快乐。利用挑高的优势做出的夹层，以悬臂式的梯板与之联系，加上夹层以透明玻璃形成半包围结构，有一种轻盈飘逸的感觉。室内的色彩也以白色和木色为主，点缀黑色的画框和个性灯饰，让空间充满艺术感。在这里生活，岂不美妙？

Hailrun
Complex
Hailrun
Complex
Hailrun
Complex

设计公司：深圳市昊泽空间设计有限公司 | 设计师：韩松 | 项目地点：中国江苏 | 项目面积：50 平方米 | 主要材料：木地板、石材、银镜、墙纸

HUAYANGNIAN REAL ESTATE WUXI XINIAN CENTER LOFT.B TYPE

花样年地产无锡喜年中心 LOFT.B 户型

Although the space is only 50 square meters, it both satisfies living functions and work area, which appears to be incredible, but is it actually realized here. Home would be like it, it does not need large space nor gorgeous style, it only needs to provide a safe haven for the mind, give you a cup of hot tea when you are cold, you can dance when your are happy, and you can stop and lick the wound when you are down. Nothing is more.

An office area is made in the entrance place, shoe racks, display cabinets, display cabinets and cabinets are used for separate space, semi-closed and semi - transparent cabinet not only ensure the independence of the space, but also ensure the space ventilation and light effect, meanwhile, the conformation based on wood is combined with wood floor and cantilevered wooden ladder board, thereby more overall sense is provided. The chairs and sofas have off-white hue, and the space is more personalized combined with the decoration of ceramics and other fashion accessories.

Hailrun Complex
Hailrun Complex
Hailrun Complex
Hailrun Complex
Hailrun Complex

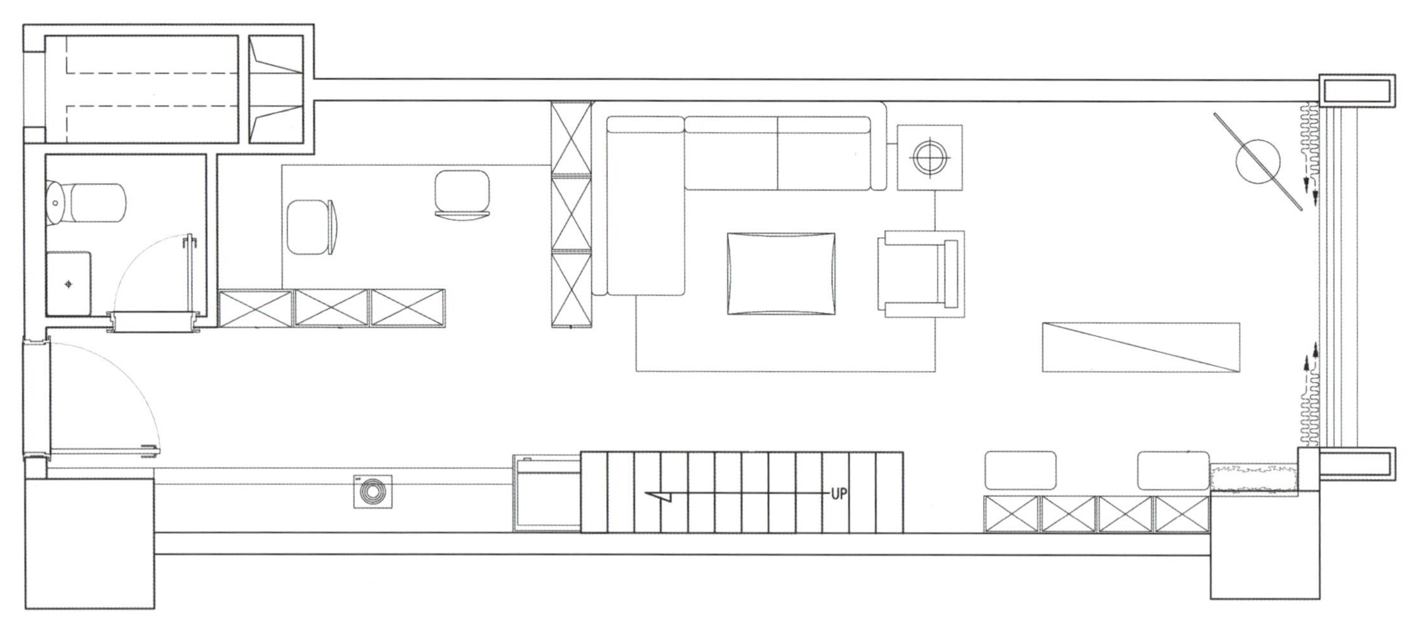
UP

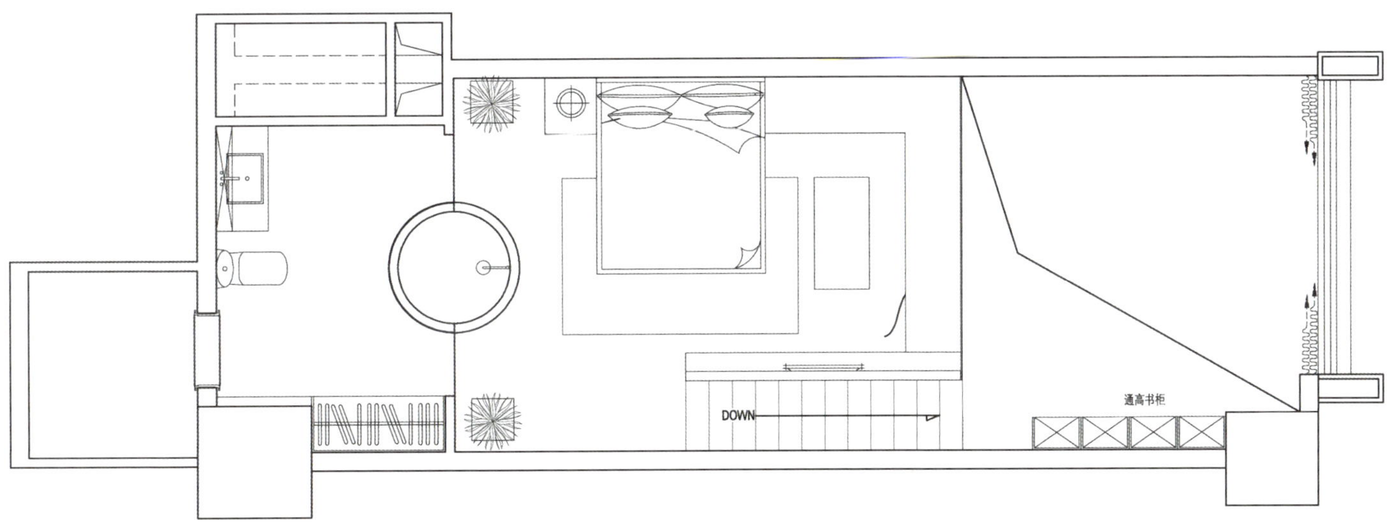
DOWN
通高书柜

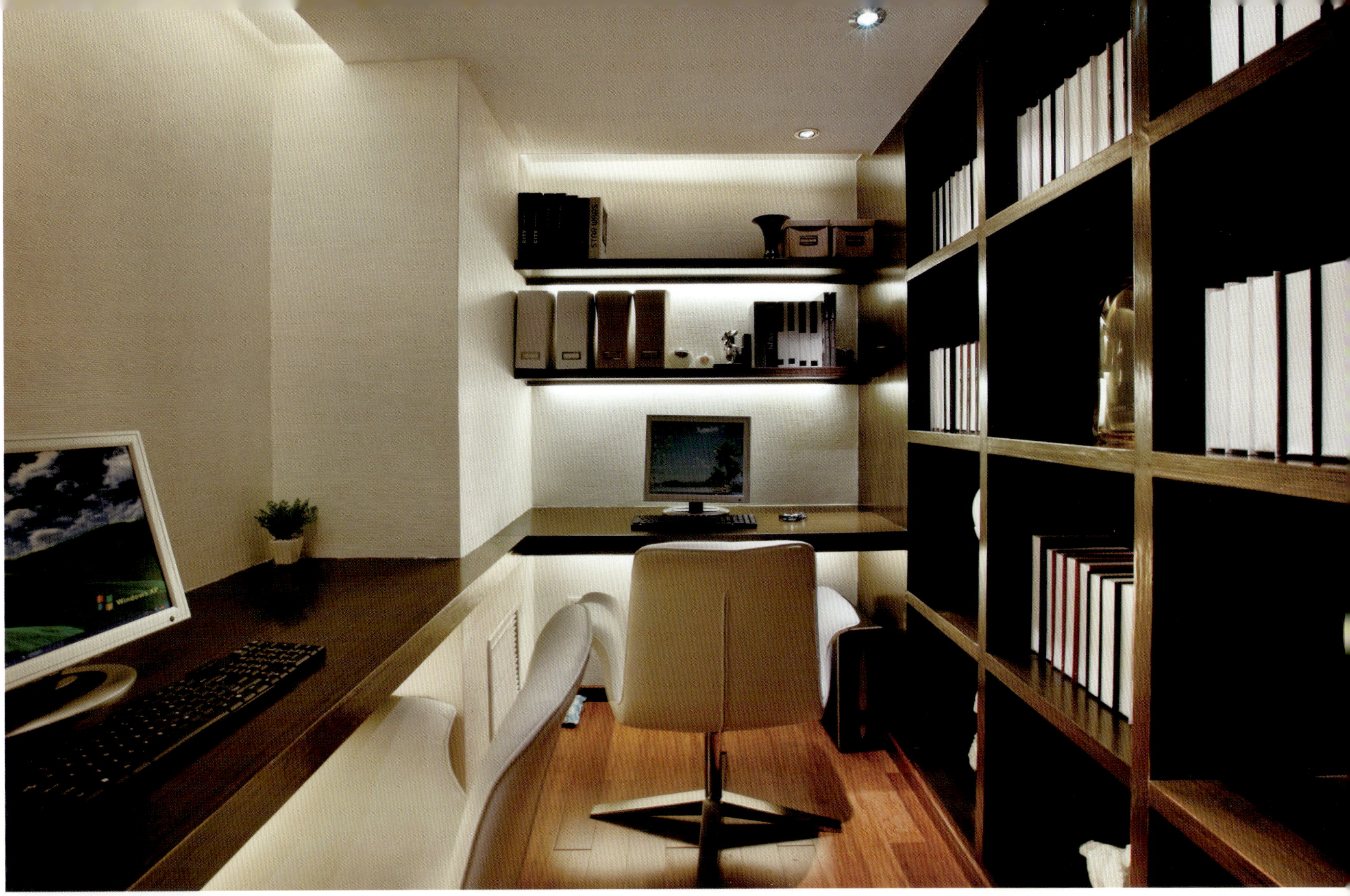

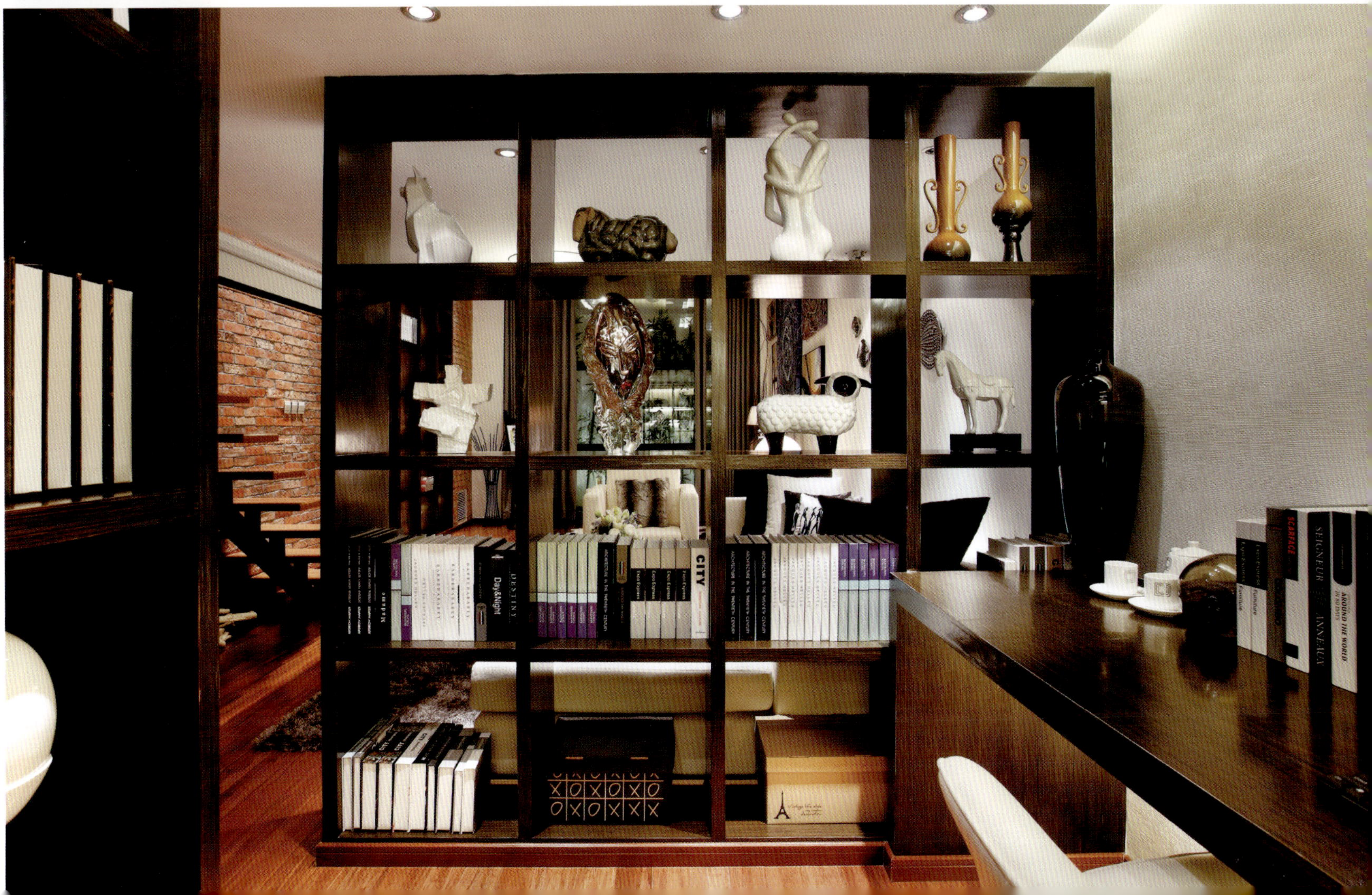

Polaroid

空间虽然只有 50 平方米，但是却在生活机能得以满足的同时，还布置了工作的区域，这看起来是一件不可思议的事情，在这里却实实在在得以实现。而且家就要这样，不需要很大，无需华丽，它只需要给心灵提供一个避风港，在你冷的时候给你一杯热茶，在你高兴的时候给你提供一个手舞足蹈的地方，在你失意时给你一个停留、舔伤口的地方。仅此而已。

本案在入户的地方做出一个办公区，以鞋柜、展示柜、陈列柜等柜体区隔空间，半封闭、半通透的柜体既保证了空间的独立性，又保证了空间通风、透光效果，同时，以木作为主体的构造与木地板和悬臂式的木梯板结合，更有整体感。座椅和沙发均采用米白色调，加之陶瓷等时尚饰品的点缀，空间更显个性。

设计公司：绝享设计工程有限公司 设计师：谢宗益 项目地点：中国台湾 项目面积：63 平方米 主要材料：大理石、烤漆玻璃、茶玻、茶镜、木皮、南方松

LIU RESIDENCE ON LONGJIANG ROAD

龙江路刘公馆

The owner is a SOHO working home, who needs studio, yoga room and cat's exclusive space besides rest, so multi-functional uses should be given to limited space. The designer displaces the toilet upstairs to create an independent large kitchen that the homeowner needs. The ceiling extends the space height and perspective sense through reflectivity of mirror stainless steel, and is cooperated with a simple table and a beautiful chandelier, and the overall space has the trend of focusing.

The super-strong incorporating stairs constructed by glass and wooden structures is light and agile through color and cross - use of irregular geometry design, the TV wall echoing with staircase is matched with the indirect light source through adopting tea mirror and wood floating cabinet, and the space sense becomes bright and general immediately. Going upstairs, the bedroom and studio with glass display cabinet and sliding door for separation adopt mobile design; thereby the space is flexible, comfortable, and structured. The simple gray low-key atmosphere brings people relaxed, comfortable and safe sense.

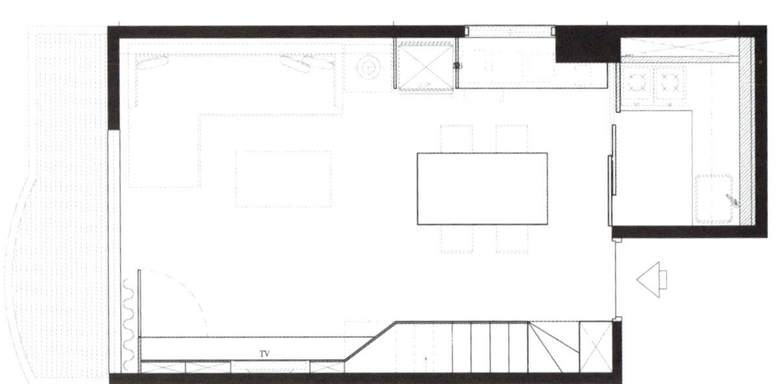
TV

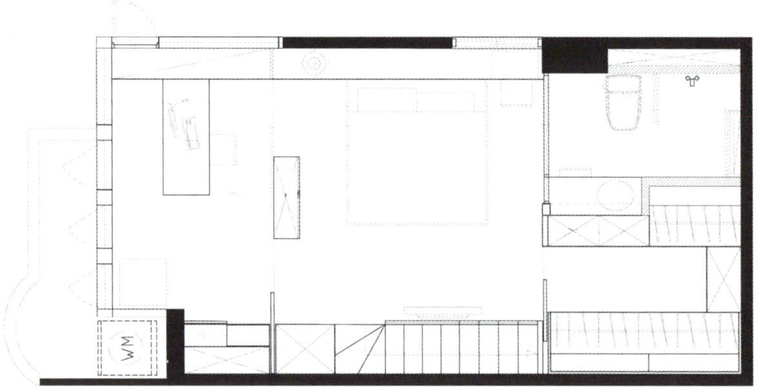
WM

屋主是在家工作的 SOHO 族，除了休憩，也需要工作室、瑜珈室、猫咪的独享空间，因此需要赋予有限的空间多功能的用途。设计师将厕所位移至楼上，打造出屋主需要的独立大厨房。天花板利用镜面不锈钢的反射性延伸空间的高度与透视感，搭配简单的餐桌及华美的吊灯，整体空间颇有聚焦之势。

以玻璃、木作构造的超强收纳楼梯，通过色彩、交叉运用的不规则的几何造型设计，显得轻盈灵动，与楼梯呼应的电视墙则采用茶镜、木作悬空柜搭配间接光源，空间感顿显明亮大气。来到楼上，玻璃展示柜及拉门区隔的卧室与工作室，采用可移动式的设计，让空间灵活、舒适，并且层次分明。而简单的灰白色以低调的氛围，带给人轻松、舒适与安全感。

ZHUBEI JINGZHAN

竹北京站

The case is narrow on the one hand, and only the L-shaped window view on the corner of the living room in the house is the lighting source of the whole house, the designer not only should solve the inherent shortcomings of short lateral width and inadequate lighting, but also should increase the admission and space function, thereby creating partitions which are separate respectively and do not avoid flow of sights, and the homeowners can enjoy ease life venue without pressure.

Designer additional arrange a loft above the restaurant, the bedroom and dressing room are mounted in the interlayer to ensure the high-ceiling of the living room. The irregular edge of the floor is matched with the semi- high railing so that the upper and lower layers can enjoy the natural light. Cabinet close to the wall is designed on the side of the corridor from the entrance to the living room, irregular shapes which set off one another falsely or truly and the cabinet door with phoenix tree weathered veneer can hide the shoe cabinet content, the rear section is used for building finished and open cabinet frame with contrast to glass and mirror materials, the cabinet frame is used for showing the mugs that are collected by the owner for many years, it also share the burden of kitchen storage.

The overall space is white with old furniture; classical style and modern style are combined in this space to create a personality and fashion.

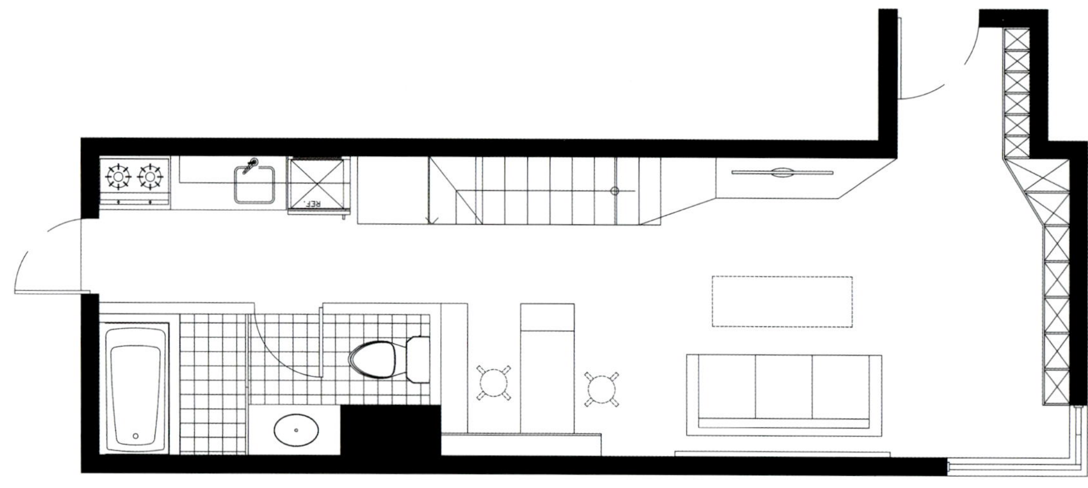
REF

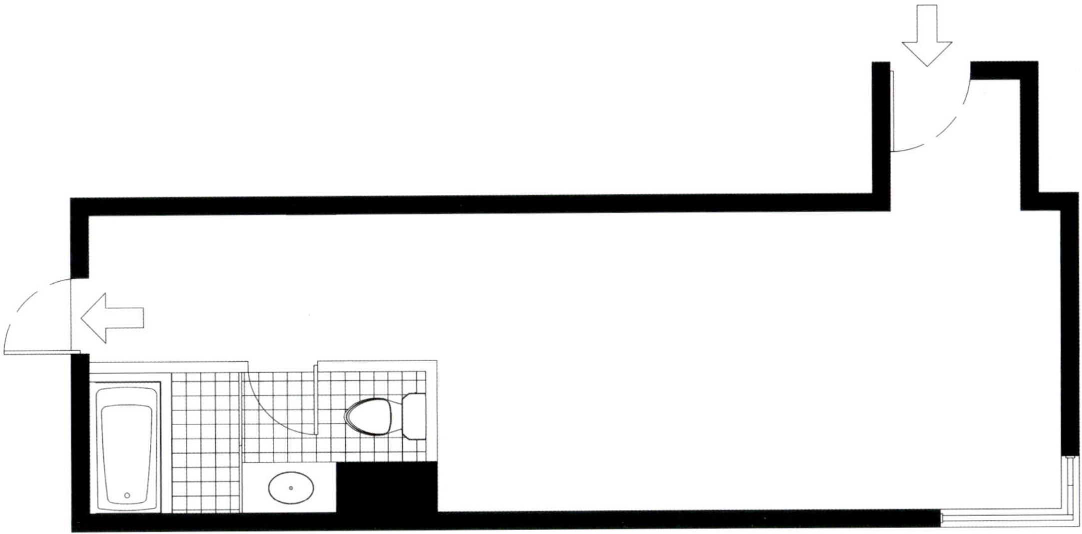

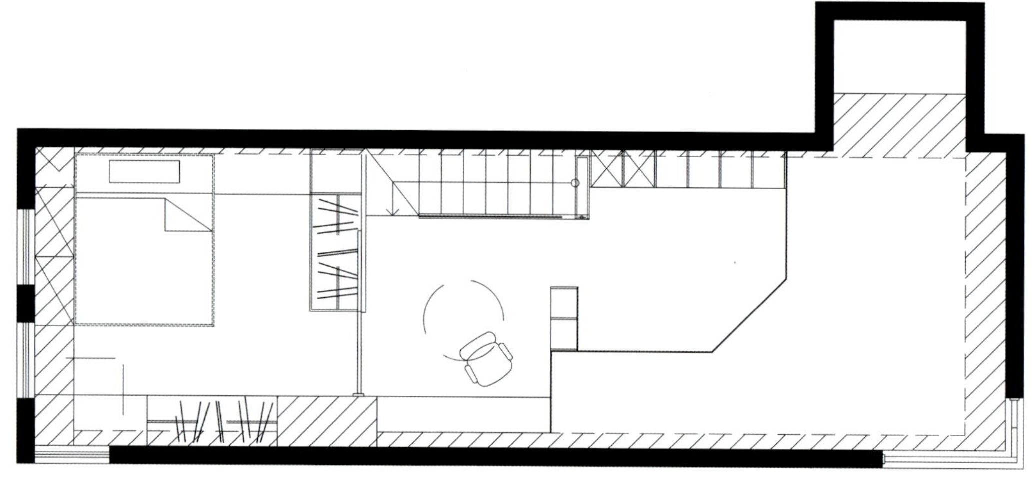

本案不但狭长，且室内仅客厅角落的 L 型窗景为全屋的采光来源，除了解决横向面宽和采光不足的先天缺点，还需增加收纳及空间机能，创造各自独立却无碍视线流动的隔间，让屋主享有自在无压的生活场地。

设计师在餐厅的上方增设阁楼，将卧室、更衣室置于夹层空间，保证客厅的挑高。不规则的楼面边缘搭配半高栏杆，让上下两层均能享有自然采光。在玄关入客厅的廊道边设计贴墙柜体，虚实掩映的不规则造型藉梧桐风化木皮的柜门隐藏鞋柜内容，后段再以对比的玻璃与镜面材质搭建利落而开放的柜架，展示屋主收藏多年的马克杯，同时也分担厨房收纳的负担。

整体空间以白色为主，带入旧家具，将古典与现代在此共融，缔造个性与时尚。

设计公司：上海塞赫建筑咨询有限公司 | 设计师：王士穌　柯津宇 | 项目面积：约 65 平方米 | 主要材料：橡木饰面、橡木复合板、橡木实木踏步、玻化砖、钢结构楼梯

XC LOFT SHOWFLAT

新城复式公寓样板房

A Loft-Studio showflat designed for Xin Cheng Real Estate. The layout is very simple: with kitchen, dining, and a double-height living space on the first floor; and a sleeping studio on the second floor. The design is minimalistic with a playful blue and white color mix. Furnishings are chic and modern. We have also incorporated some unique items such as old cameras, black and white photos, mountain bike etc. to accentuate the character of the house owner.

为新城地产设计的小复式户型，主要针对年轻客户群体。整体设计非常简洁：厨房、卫生间、餐厅、客厅位于一楼，卧室与工作室位于二楼。设计尽量保持户型的通透性以及明亮宽敞的氛围，并采用了个性感较强的白色＋蓝色的整体色调搭配。软装方面除了精致的家具以及墙纸之外，我们也利用了老相机、照片、脚踏车等特殊饰品来体现屋主的个性以及生活氛围。重点体现户型可成为年轻业主秀出个性的展示宅体。

设计公司：研宇整合设计 | 设计师：吴建宏 | 地点：中国台湾 | 面积：39 平方米 | 主要材料：木皮、黑镜、超耐磨地板、玻璃、盘多磨地板

XINYI ROAD

信义路

The designer makes full use of the space advantage of 4.2 meters height, and makes an attempt to maximize this limited space, to gain more space for residents. The designer consolidates space functions, sets the sandwich construction above open-style kitchen and the porch, and makes this living space where people stays there commonly for a long time show spacious and magnificent atmosphere by means of this high space advantage. This special steel sandwich construction, plus a sense of passing through transparent tempered glass make this existing sandwich construction no pressure, while the living space reserves large area French window to take in outdoor scenery to the house.

The house owner himself collects thousands of CDs, which must be planned and placed in order. The designer makes full use of the space height advantage and plans high cabinets on both sides of TV wall in the living room, where it is convenient to fetch and place the iron ladder. The designer skillfully makes use of entryway of stairs, and sets the storage cabinets with different sizes, which makes no effect on daily work and rest, and resolves the difficulty of not being beautiful for this storage function as well. This deposit box is hided in the shoes cabinet, which is closed to be a porch display counter. There are two functions about this little cabinet, with no trace or being invisible to store items.

+159.6
+182.4
+136.8
+114
+91.2
+68.4
+45.6
+22.8
H=308
H1:181
H2:119

156X190
+159.6
+182.4
+136.8
+114
+91.2
+68.4
+22.8
H=308
H=316
H1:181
H2:119

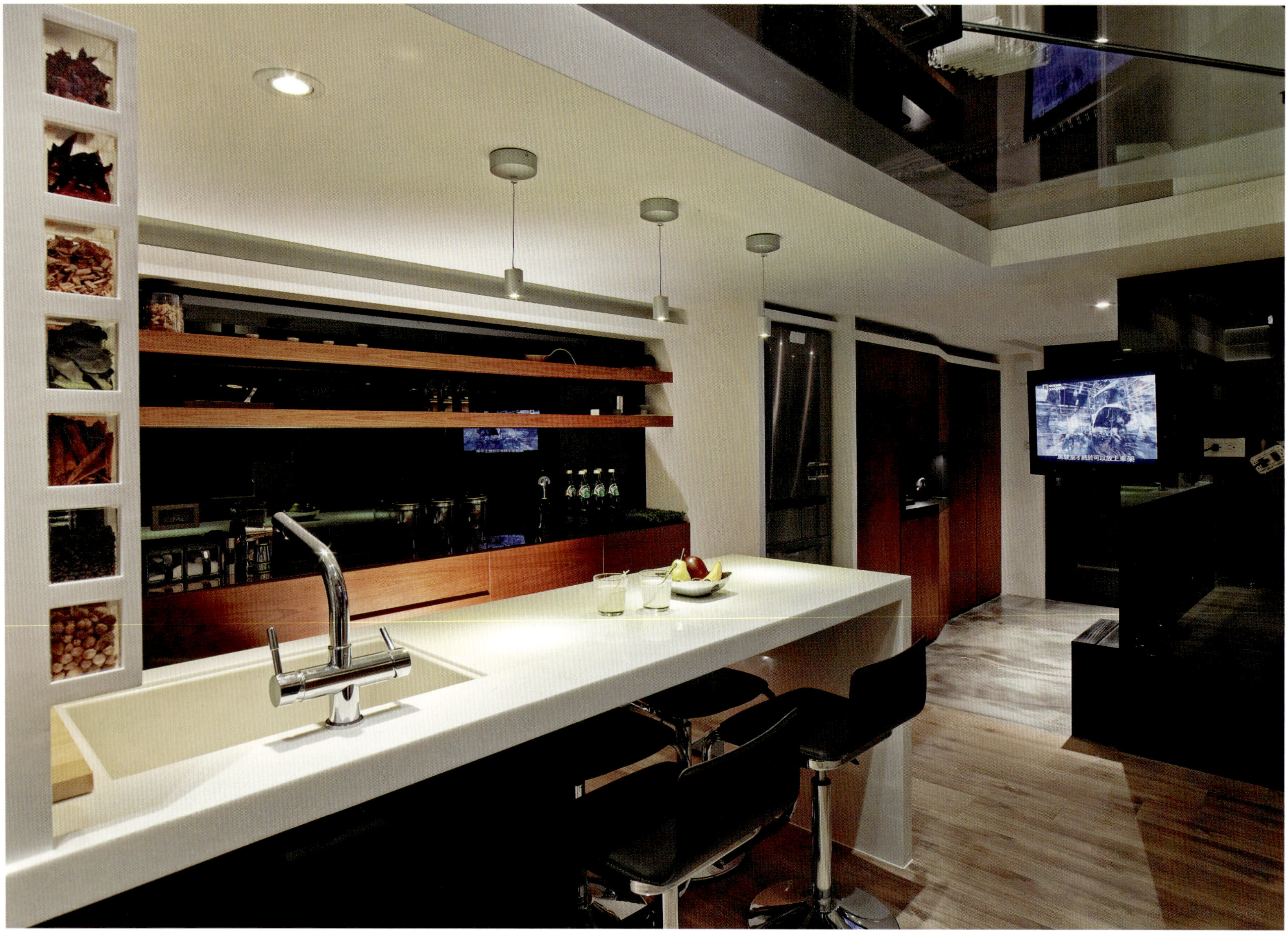

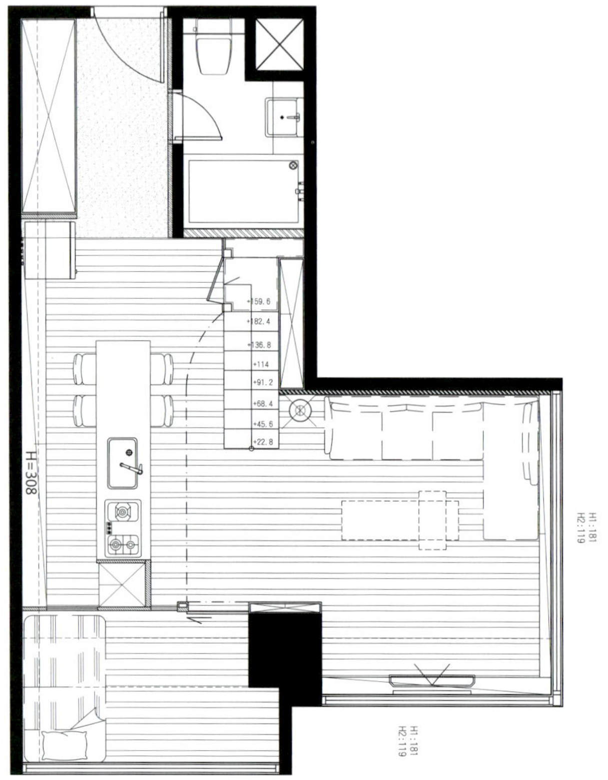

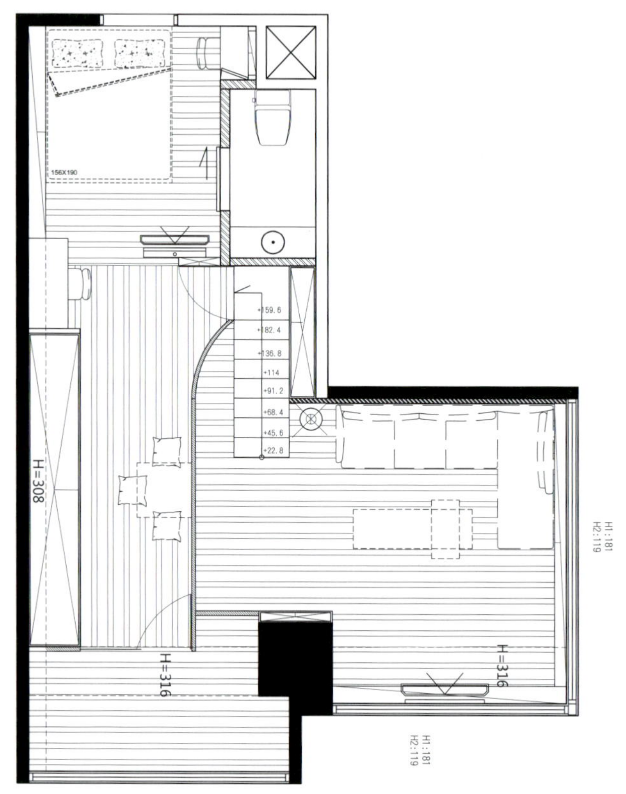

设计师利用空间挑高 4.2 米的优势，设法在有限空间内发挥最大效益，为居住者争取更大使用空间。设计师整合了空间的功能性，将夹层设置在开放式厨房与玄关上方，始较长久待的客厅空间藉由挑高优势，展现宽敞大气氛围。而特殊的钢构夹层，加上透明强化玻璃的穿透感，让夹层的存在完全没有压迫感，客厅空间则保留大面积落地窗，将窗外景致收入屋内。

屋主本身收藏近千片 CD，必须有条不紊的规划摆放，设计师利用挑高优势，在客厅电视墙两侧规划高耸的柜体，搭配铁梯方便取放。活用属于过道空间的楼梯转角，设置大小收纳柜，既不影响平日作息活动，也解决收纳不美观的既定形象，连保险箱都藏在鞋柜之中，合上时就是玄关展示台，一个小机关两样用途，不着痕迹收纳于无形。

设计公司：马迪思．明团设计机构 | 设计师：Manuel Derel、Eric Wong、Julian Cornu、Philippe Colin | 项目地点：中国广东 | 项目面积：67.4 平方米

WAN JIA ZHENG HAO GOLDEN PALACE FIRST PHASE 18#

正佳万豪金殿一期 18#

This project belongs to a small dwelling-size duplex. In this not large space, the space presents to be vivid and bright because of a mezzanine. Getting out of the area limitation, the modern and fashionable style going with new classical elements, this small space shows a different temperament of a luxury real estate.

Because of the space limitation, the designer regards the whole first floor as the public space to build, where the open kitchen, passageway, the dining room, the living room and the bath room can be represented perfectly as the most basic life functional area. The kitchen adopts black going with white, and there is a distinct and personalized scene built at the entrance. The passageway from the kitchen to the dining room is the outside of kitchen, even two elegant and unique European seats offer people a leisurely and tea-drinking space. The living room and the dining door adopts pure white design, adorned by a tinge of black furnishings, plus the higher position advantage of the living room and the design of the whole glass wall, offering people an elegant and not realistic feeling.

BUILDING
LONDON

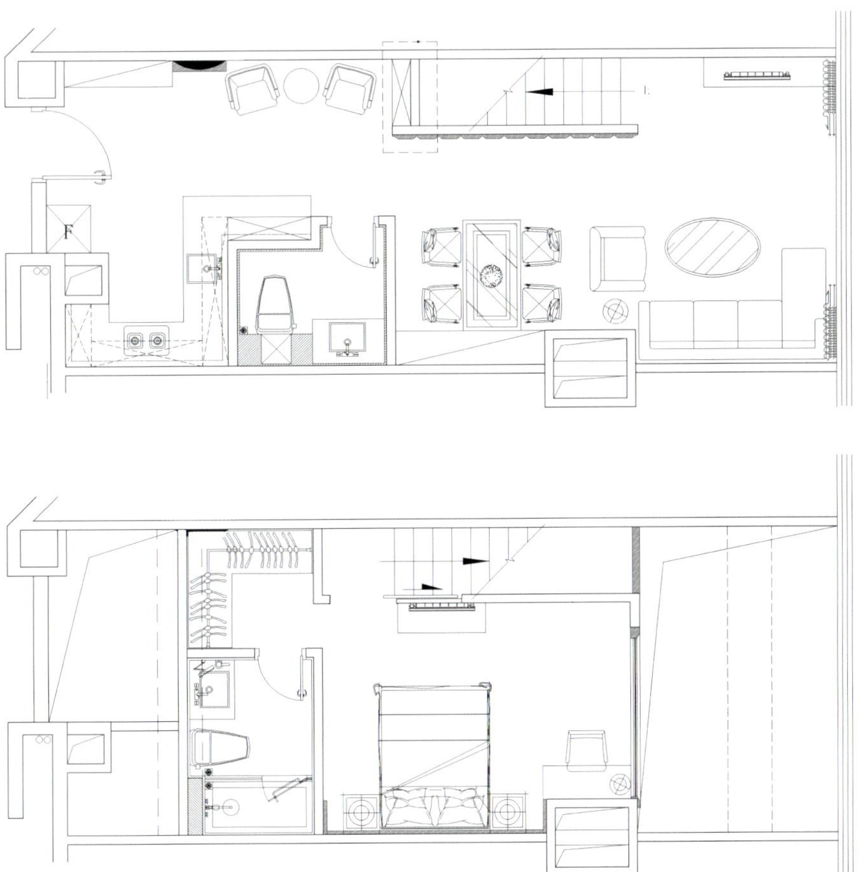
F

本案属于小户型复式结构，在面积不大的空间里，因为多了一个夹层，空间便显得灵动开朗起来，挣开面积上的局限，以现代时尚风格搭配新古典元素，小空间也呈现出不一样的豪宅气质。

因受空间局限，设计师将整个一层全部作为公共空间来打造，开放式的厨房、过道、餐厅、客厅以及卫生间，将最基本的生活功能区完美呈现。厨房采用黑白色搭配，在进门处营造一处独特个性的风景线。在厨房通往餐厅的过道处，也就是卫生间的外面，甚至还摆放了两张造型优雅独特的欧式座椅，给人提供一个休闲品茗的去处。客厅和餐厅采用纯白色设计，点缀些许的黑色陈设，加上客厅的调高优势和整幅玻璃墙的设计，给人一种晶莹飘逸的不真实感。

设计公司：界阳 & 大司室内设计 | 设计师：马建凯 | 项目地点：中国台湾 | 项目面积：66 平方米 | 主要材料：镜面不锈钢、灰镜马赛克、LED 灯、胶合强化玻璃

GUEST HOUSE WITH FASHIONABLE NIGHTCLUB STYLE

时尚夜店风招待所

The case belongs to duplex apartment in prime location; the homeowner adopts the room to entertain friends and customers as private gathering place, and specifies the design theme of 'Fashionable Nightclub Style'.

The designer created the case with very fashionable design style. Tailor-made furniture is used for dealing with and conciliating irregular pattern formed by combining two high-ceilinged suites with different sizes, the designer specially made the size of the sofa and L-shaped angle so that furniture can closely fit the twisted wall of the living room, yellow light is produced from the lower portion of the gray mirror drawer, thereby increasing the light atmosphere of the night hours. Interlayer treatment is made for the 3.6 -meter-high ceiling, the upper and the lower height is slightly less so the designer hollowed up the floor partially, thereby boldly making the block transparent with visual penetration, the sense of space can be drawn, and the most attractive focus of the bar is creased at the same time.

The entire space is filled with a colorful sense of dynamics, and strong visual concept is boldly brought, thereby creating nightclub atmosphere that people exclaimed again and again in every step.

本案属于黄金地段的跃层户型，是屋主用来招待朋友、客户的私人聚会场所，并且指定了“时尚夜店风”的设计主题。

设计师以极具时尚感的设计风来打造这一案例。对于两户大小不一的挑高套房合并形成的不规则格局采用特制的家具来处理、调和，设计师特制了沙发尺寸与 L 型角度，使家具能紧密贴合歪七扭八的客厅墙面，并在灰镜抽屉下方打黄光，增加夜晚时分的光影氛围。对于 3.6 米高的跃层做夹层处理，上下的高度略显不足，于是设计师将楼板局部挖空，大胆地将阻碍透明化，使视觉穿透，以此拉高空间感，同时打造出吧台最亮眼的焦点。

整个空间充满了五光十色的动态感，大胆带入强烈的视觉概念，打造出每一步都令人惊呼连连的夜店氛围。

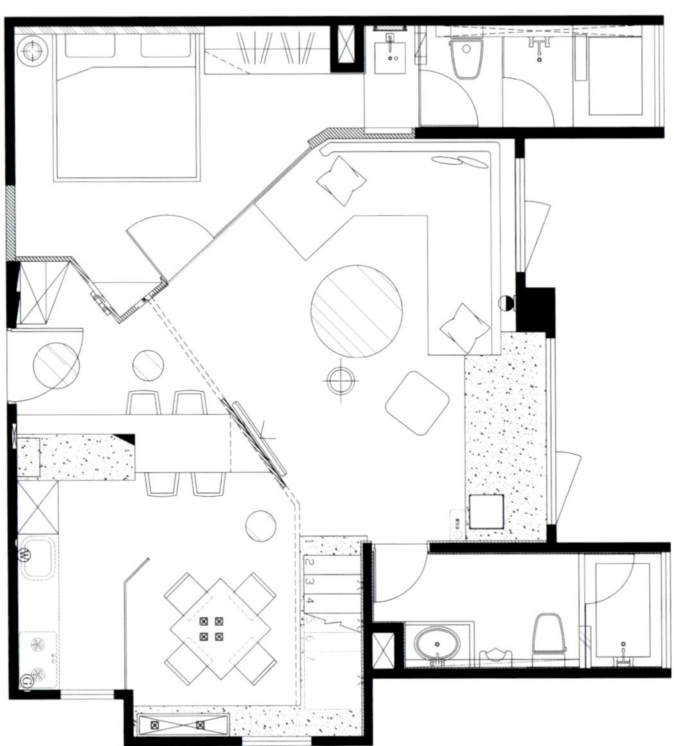

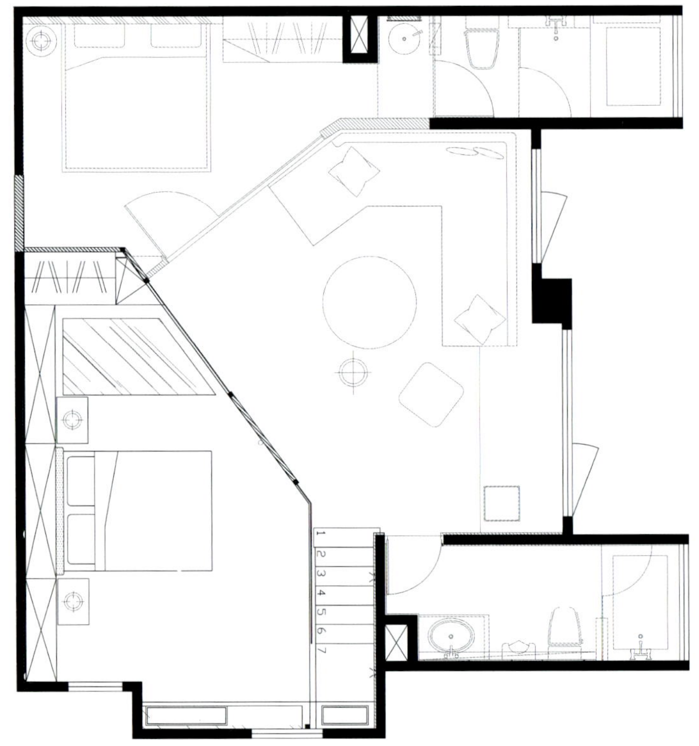

设计公司：界阳&大司室内设计 | 设计师：马建凯 | 项目地点：中国台湾 | 项目面积：72.6 平方米 | 主要材料：文化石、钢刷梧桐木、银狐大理石、舒然板、LED 灯

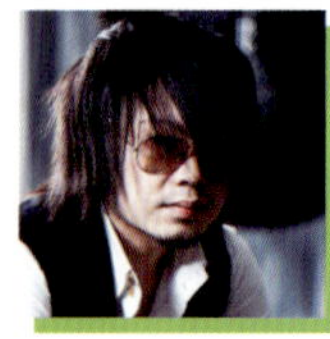

MISS RICE

北投赖小姐

The designer breaks the traditional design practice, reserves the current irregular house profile, and plans inclined lines to cooperate with the mezzanine, in order to brighten the first floor through multidimensional lightings and pass through the clear glass armrest on the second floor, which goes against traditional design lines and a humanistic sense of wood materials, and mixes into a new humanistic lighting space.

The huge superior wooden dining tables and chairs are placed in the center of the space, a sense of existing feeling which cannot be ignored. The designer adopts the dining table as the center, and plans the separate district function along the wall in the open space. On the right of the porch shows the storage shoes cabinet, ahead of which there are the custom-made sofa in the living room and the transparent bubble lamp dropping from the ceiling of 4.2 meters, regulating the living room atmosphere combined with illusory cultural stone pillars. The storage room, the living room and the bath room are on the both sides of the bridge on the left side of the porch, moving line windingly extending forward, so the designer uses water blue baking paint to plan L-shape dining bar. Moving lines extending to the left and the right meet at the bed area in front of windows nearby the dining table, and the low platform introduces abundant lightings, which not only can be employed as the seats for guests gathering, but also is the most comfortable sunlight reading area.

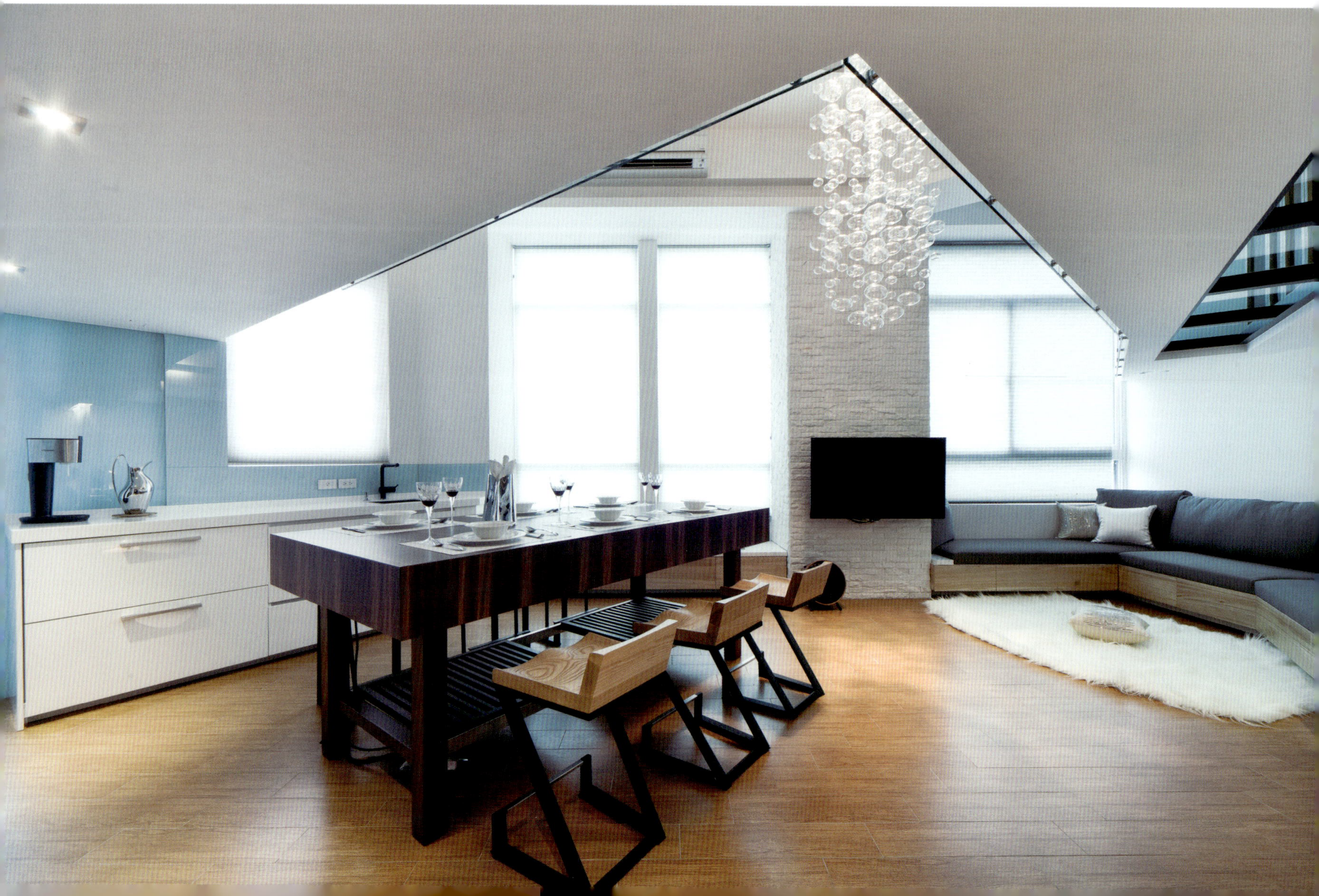

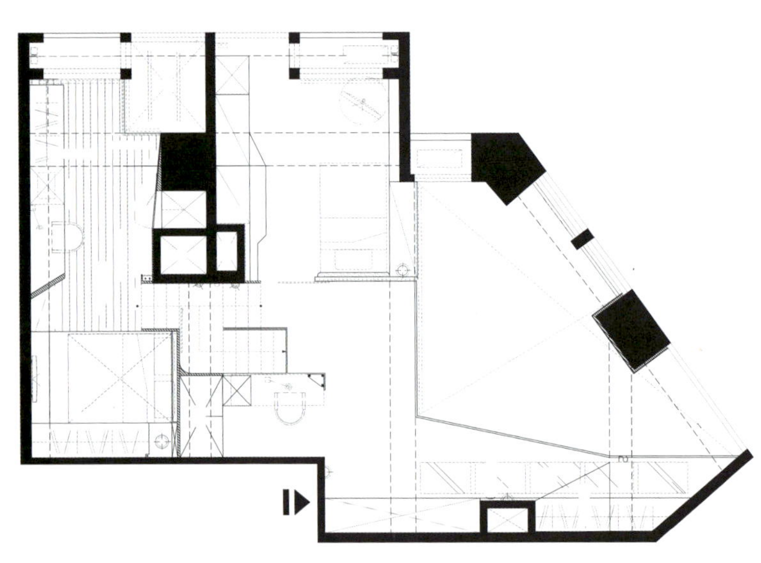

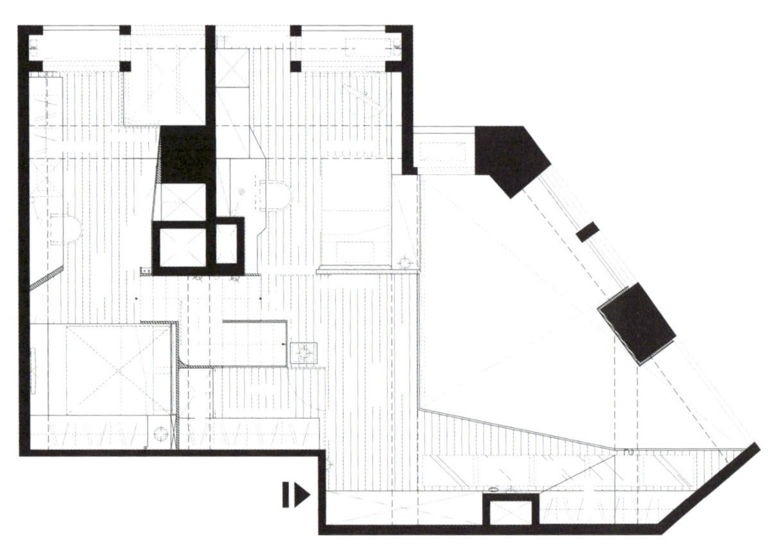

设计师突破传统的设计常规，保留现有不规则的屋廓，并在夹层处规划与之呼应的斜切线条，让多面采光照亮一楼，也穿透二楼的清玻扶手至后方的卧室，颠覆传统的设计线条与人文质感的木作，交融一场新人文日光空间。

偌大的高级木质餐桌椅置于空间正中央，有着不容忽视的存在感。设计师以餐桌为中心点，在开放空间里沿墙规划独立区域机能。玄关右边是展示收纳鞋柜，往前是订制的客厅沙发，透明气泡灯从四米二的天花板悬垂而下，与虚化柱体的文化石柱定调客厅氛围。玄关左边，冰箱两侧分别是储藏室及客卫，动线向前延伸转折，设计师以水蓝色烤漆依照窗型规划 L 型简餐吧台，左右延伸的动线交会于餐桌旁的窗前卧榻区，压低的台度引进充足采光，不仅可作为宾客齐聚时的座椅，亦是最舒适的日光阅读区。

设计公司：绝享设计工程有限公司 \ 设计师：陈冠樵 \ 项目地点：中国台湾 \ 项目面积：39.6 平方米 \ 主要材料：抛光石英砖、镜面铝板、白色美耐板、超耐磨地板

BANQIAO-NEW DOME

板桥——新巨蛋

Small unit space often gives people an impractical feeling mainly due to insufficiency of storage function and regional unclear boundaries. It can cause the embarrassment of seeing through the whole room though it can create a visual extension effect due to lack of facade barrier coverage, thereby the designer should control well between sense of permeability and privacy.

It is primarily necessary to strengthen the storage functions in order to avoid loss of home practicality due to limitations of area. Symmetric closed -door cabinets are arranged on two sides of the TV wall in the living room area, indirect lighting is used for reducing the somatosensory below the cabinet, a row of drawers are also designed to contain piecemeal items, the drawer table also can be used for carrying audio-visual equipment at the same time, and two functions can be met with one design.

The public space on the first floor is divided into the living room and tatami room, and tatami room is not provided with separate gate, and is only divided by elevated wooden floor. As a result, the broad sense of space is retained to make the tatami room to become extension of the living room on one hand, the embarrassment brought by blurred borderline sense is also avoided on the other hand at the same time.

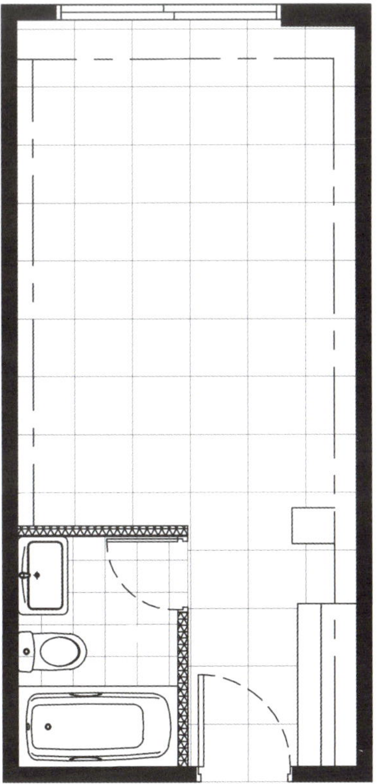

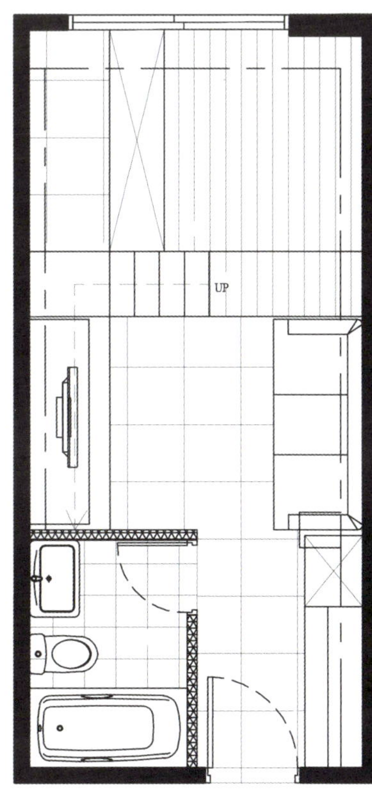
UP

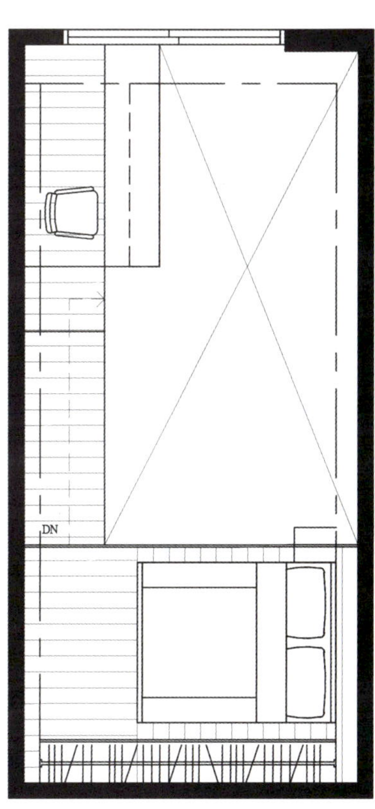
DN

小户型空间经常给人一种中看不中用的感觉，主要是收纳机能不足和区域界定不明所致。少了立面屏障的遮掩，虽能营造出视觉延伸的效果，却也会产生一眼看透全室的尴尬，因此要在通透感与隐私间拿捏好分寸。

为了让住家实用性不因面积的局限而有所缺失，强化收纳功能成为第一要务。客厅区在电视墙两侧安排了对称的封闭型门柜，柜体下方除了用间接照明来轻化量体感，还规划了一排抽屉柜归纳零碎物品，抽屉台面同时成为影音设备的置放处，用一种设计就满足了两种机能。

一楼的公共空间划分为客厅及和室，和室不另作门片，仅通过架高的木地板作为分野。这样一来，不仅保留了空间的开阔感，让和室成为客厅的延伸；同时也避免了界线模糊带来的困窘感。

Simmons

设计公司：绝享设计工程有限公司 | 设计师：李为琢 | 项目地点：中国台湾 | 项目面积：16.5 平方米 | 主要材料：喷砂玻璃、胶合玻璃、白膜玻璃、美耐板、喷漆

INNATE WINNER

天生赢家

Many beams structure can be observed in a high ceiling space, the designer used the space under the beams to create an entrance shoe cabinet, the simple and elegant white cabinet and kitchen are combined to create an overall sense, a mirror is matched in the middle, and keys and small objects can be placed in the inner concave place.

The best place in the house is a large balcony, thereby the designer retained the entire floor to ceiling window, the light is projected inside completely to create a fresh and bright atmosphere coupled with the design of the shutter. No interval is available between the kitchen and living room, the designer utilized the wooden cabinet to separate space. An entire dressing mirror is made on the side in face of the entrance door, and the design of the toilet utilized matte stainless steel to combine with sandblasted glass door, which is both simple and light-catching.

The stairs were made in the space above the TV cabinet in the corner of the house, the design of combining with the TV wall saved more space. The sandblasted glass on the second floor reveals a hint of green with refreshing and pleasant feeling, which has a feast for the eyes as well more fashion sense.

挑高的空间看得到许多梁柱结构，设计师利用柱子下方的空间做了玄关鞋柜，简单素雅的白色柜体与厨具结合营造出一种整体感，中间搭配明镜，内凹的地方可放置钥匙与小物。

房子最棒的地方是有着很大的阳台，因此设计师保留了整面的落地窗，让光线完整地投射进来，加上卷帘的设计，营造清爽、明亮的氛围。厨房与客厅之间没有间隔，设计师利用木作柜体区隔空间。面对玄关大门的侧边则为业主做了一整面的穿衣镜，而厕所的设计则利用雾面不锈钢搭配喷砂玻璃门，既显简约，又可采光。

在房子的转角处利用电视柜上方的空间做了楼梯，与电视墙的结合设计更加节省空间。二楼的喷砂玻璃透着淡淡绿色，有种清爽怡人的感觉，除了赏心悦目，更添时尚感。

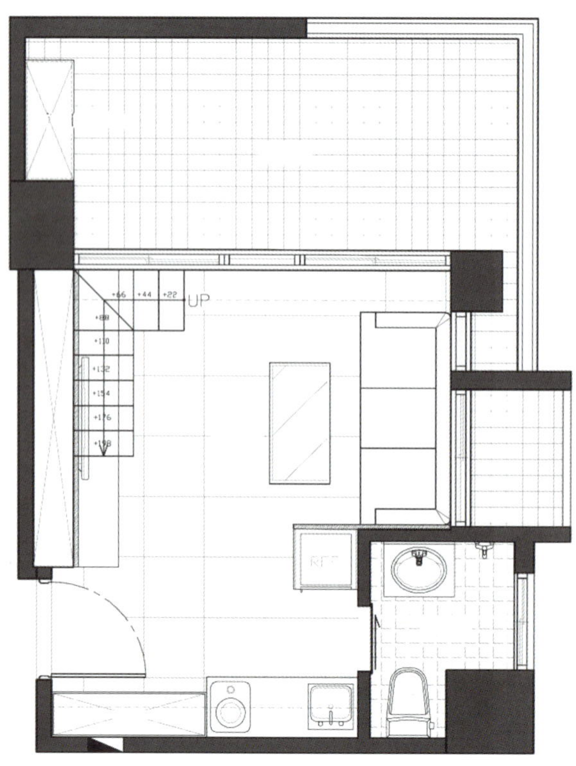
UP

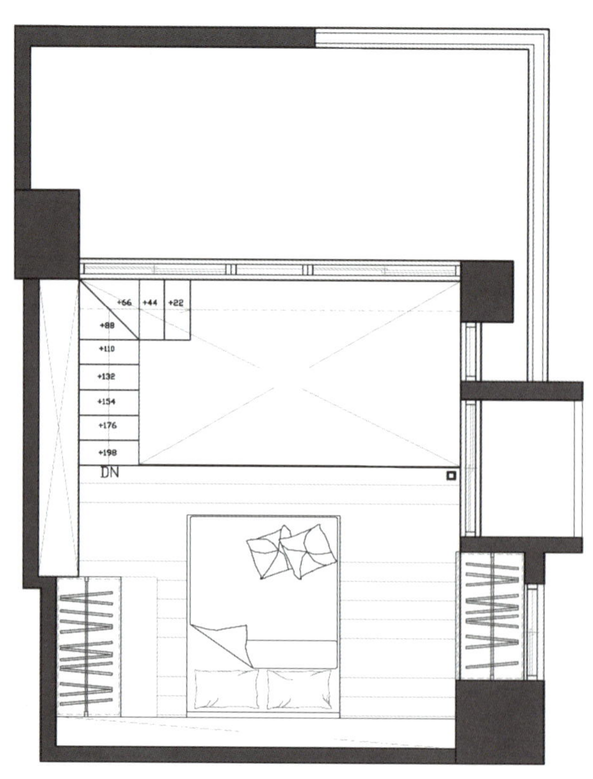
+44
+22
+88
+110
+132
+154
+176
+198
DN

设计公司：台北拾雅客空间设计 | 设计师：许炜杰 | 项目地点：中国台湾 | 项目面积：59 平方米 | 主要材料：银狐大理石、壁纸、进口磁砖、玻璃

DRIVING DESIGN MOVING FACTORS AND EVOLVING FURTHER PERFECT LIFE

驱动设计动人因子 进化下一步完美生活

Stereo model ceiling reflects the light of diamonds in the sun shining, stereo lines after deformation magnify the visual space, crowded broken sense brought by excessive function pattern segmentation can be removed. The large bar planned for the female owner with the favor of baking western cookies cleverly designs the walking line of the porch, and gradually enlarging sense of space can be brought. Unimpeded sunlight window scene extends from the TV wall of the living room, thereby every open zone can enjoy natural clear light, and the space vision is more amplified.

The designer uses 3.45 m ceiling advantages for planning interlayer space above the study, thereby creating a guest room. The relations of the upper and lower layers can be lined with the shaped layer in fresh yellow in net white, the space under the stairs is utilized to design the storage room for kitchen and half bath, the stair located in the center of the space can exert the function to the maximum. The mater bedroom on the last section can be elastically adjusted into open or independent life function according to living member changes, thereby further improving life.

立体造型天花在日光的映照中折射出钻石般的光芒，变形后的立体线条放大了视觉空间，消弭了过多的机能格局分割带来的拥挤零碎感。设计师为喜爱烘焙西式小点心的女主人规划的大型吧台则巧妙地规划出玄关的行走动线，带给人渐行放大的空间感受。无阻碍的日光窗景从客厅电视墙处向内延伸，让每一个开放区域都享有自然敞亮的光照，更加放大了空间视野。

设计师利用 3.45 米的挑高优势，在书房上方规划夹层空间，做出一个客房。在一片净白中以鲜黄色的造型层架链接上下层的关系，并利用梯下空间规划厨房用储藏室及半套卫浴，将位于空间中央的楼梯功能发挥到极致。最后一段的主卧室，可随居家成员的变化弹性调整为开放或是独立的生活机能，进一步完善生活。

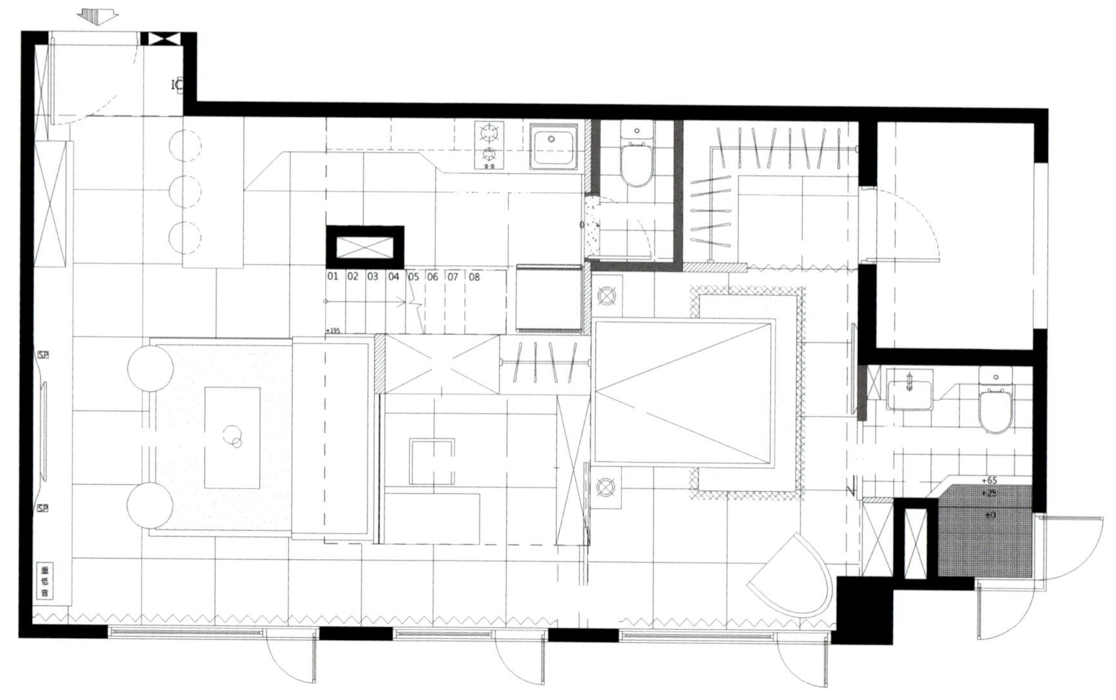

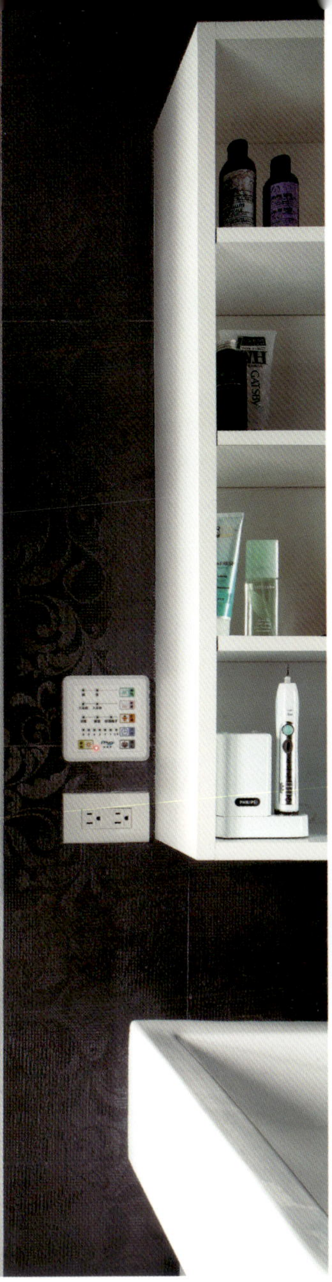

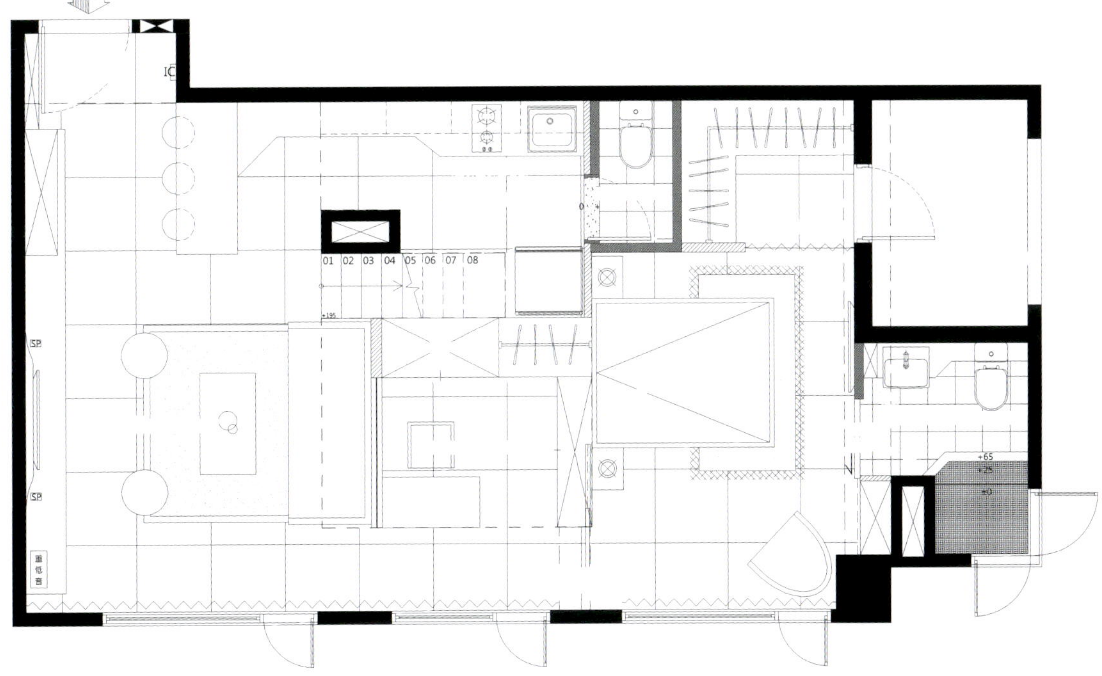

设计公司：台北拾雅客空间设计 设计师：许炜杰 项目地点：中国台湾 项目面积：66 平方米 主要材料：烤漆玻璃、玻璃马赛克、不锈钢

LINES STREWN AT RANDOM SHOW SPACE AMPLIFIED DOUBLY

错落线条 玩出双倍放大空间

Designer creates style features only belonging to the residence according to the life characteristics of dwellers through arrangement by the moving lines, and customization of materials, furniture and soft packs, the value of the space itself can be amplified, diversified, more active and more fulfilling through interactive relation of penetration, level, extension, duration, contrast, proportion and external link through unified consonance of design.

Each separate space has a light side in readjusted pattern, thereby introducing daylight to enlarge the space. If we look backwards from the half-height living room TV wall, our vision can reach the rear main bedroom beyond the working small platform, the ladder compositely designed upwards extends from the dining room located in the center of space axis, and enters the girl room through the turning of the small platform, the 3.4 -meter-high façade can be well used, and the sense of space is created with the staggered use of real wall and clean glass.

The geometric elements with strong sense of lines are introduced to weaken the sense of pressure from the oblique angle in the shaping of the overall space, thereby improving the extension of the vision and shaping mansion -like momentum.

DESIGN

设计师根据居住者的生活特质，通过动线安排、材质、家具、软装的定制化，打造宅邸专属的风格特色，并将穿透、层次、延伸、持续性、对比、比例及对外连结的互动关系，透过设计的统一调和，让空间本身的价值放大、多元化，也更积极、更充实。

重整后的格局中，每一个独立空间都有采光面，引入日光放大空间。从半高客厅电视墙向后方望去，视野可越过工作小平台直抵后方的主卧室，而向上复合设计的楼梯从位于空间轴线中心的餐厅处延伸，经过小平台的转折向上进入女孩房，善用 3.4 米高的立面，并以实墙与清玻的交错运用来塑造空间感。

在整体空间的塑造中，通过导入线条感强烈的几何元素，以斜切角度弱化压迫感，提高视野的延伸度，形塑大家豪宅般的气势。

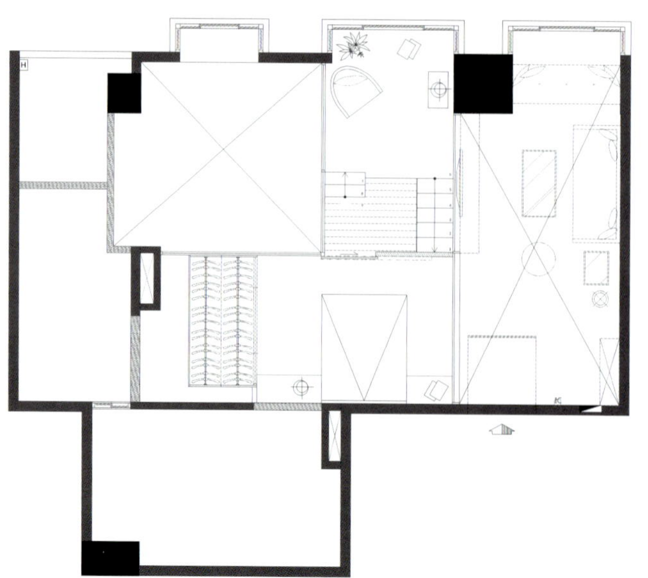

设计公司：隐巷设计顾问有限公司 | 设 计 师：袁筱媛 | 项目地点：中国台湾 | 项目面积：50 平方米 | 参与设计：LUNA | 摄影：王基守

主要材料：锈蚀黑铁板、超白强化烤漆玻璃、12mm 强化清玻璃、黑色烤漆玻璃、冲孔铝板、60mm 拉丝不锈钢管、3mm 钢索、文化石、黑板漆

ZHONG MANSION ON SHUIYUAN ROAD

水源路钟邸

The owner couples in this case are flat creative workers, since creativity and life are inseparable, additionally, it extremely difficult to find a house with breathing and feeling in the metropolis, and thereby they decided to start a new life in the living district in Hedi Park, and thoughts and creativity can carry out photosynthesis through water scene and lawn. The space has lengthwise direction, which is similar to standard room of star hotel, space has longitudinal length of 3.6 meters with a window to the south.

Considering to the working industry of the owners, the designer proposed the home concept of the work combination style in home, and retained the home's basic functions, such as kitchen, living room, bedroom and the like. The dining area and work area are combined, and the desks and chairs on the first floor can be movably combined and used elastically according to demand.

The entire space looks like a creative production base, all kinds of old furniture are combined with creation integration, and thereby practicality and beauty are exerted to the maximum. The close integration of creativity and life shows the self-thought of the owners, namely home with a complete studio.

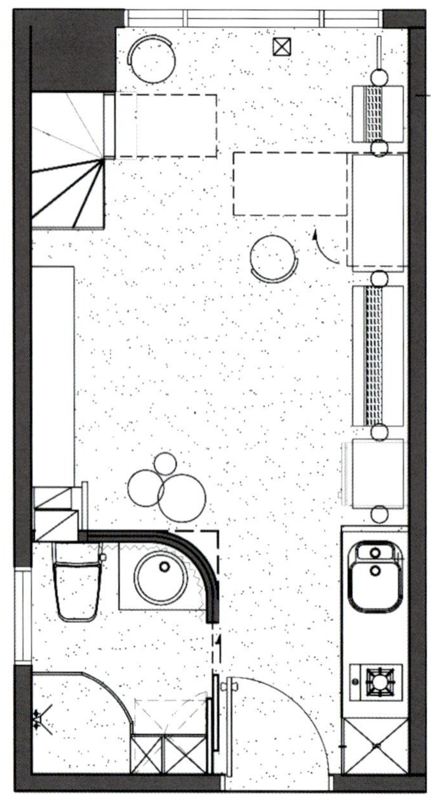

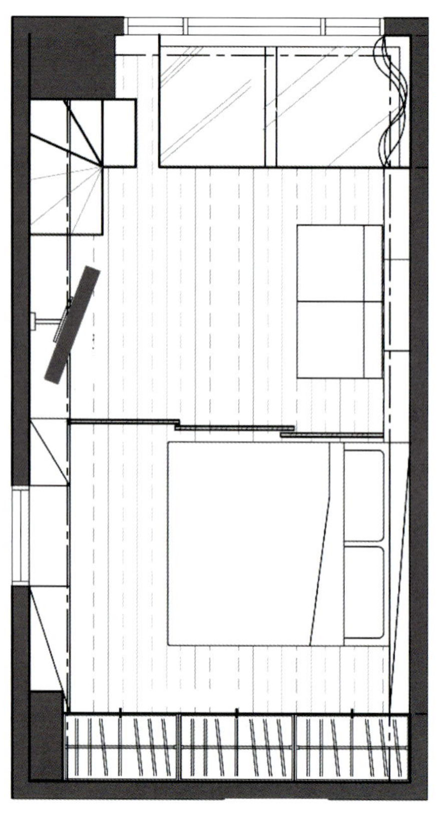

K·SPACE
58

IGHT THE FIGHTERS
NOT
THEIR WARS

本案的业主夫妇为平面创意工作者，因创意与生活密不可分，加上在大都会中寻找会呼吸、有感觉的房子如大海捞针，于是决定在有河堤公园的小区展开新的人生，透过水景、绿茵让思绪和创意进行光合作用。空间为纵长向，类似星级饭店的标准客房，空间小挑高 3.6 米，南向面窗。

考虑到业主的工作业态，设计师提出结合工作式的家的概念，在不大的居住空间中，保留住家的基本功能，如厨房、起居、卧室等。餐厅区域与工作区结合，一楼的桌椅皆为活动组合，可依据需求弹性使用。

整个空间就如同一个创意生产基地，各式各样的老旧家具搭配创意组合，将实用与美观发挥到极致。创意与生活的紧密结合，展现出业主自我的想法，圆满工作室的家。

YOU ARE THE

58
YOU ARE THE

58
YOU ARE THE

102–225

REFINED SMALL DWELLING-SIZE APARTMENTS

精致小户型

设计公司：武汉多维空间艺术有限公司 设计师：张晓莹 陶清明 熊丹 项目地点：中国山西 项目面积：50 平方米 主要材料：木作、白漆

TIANSEN PEACE NO. 1 SHOW FLAT

田森安宁一号样板房

She is a single woman of 35 years old.

In this 50 square meters' space, bathtub can be finally liberated from a small private space and become the main subject of her room, proudly and freely.

In the opposite side of bathroom is her favorite wardrobe, where we can find her beloved " living artwork " releasing their own light like their owner, but no more hidden in a closed wooden cabinet, because they are brought to be enjoyed but not to be forgotten.

She don't need to cook very often, but delicious food is still obviously one of a woman's favorite things. Reword yourself a good meal occasionally, and then "Cooking" and "Eating" are no more ruled or divided "beautifully", you can just do as you like and enjoy yourself.

The living room is connected with the dinning place, making perfectly a multifunctional space, where she is able to enjoy her meal lying on the sofa, or to watch soap TV series sitting besides the dinning table.

study room is connected with living room in the most delightful place, and it's also possible to be separated and become a individual space which can satisfy her occasional desire of writing something□

Her bedroom is towards the desk, this is a place where can be relatively private. Because She still needs to feel comfortable and in security when she sleep.

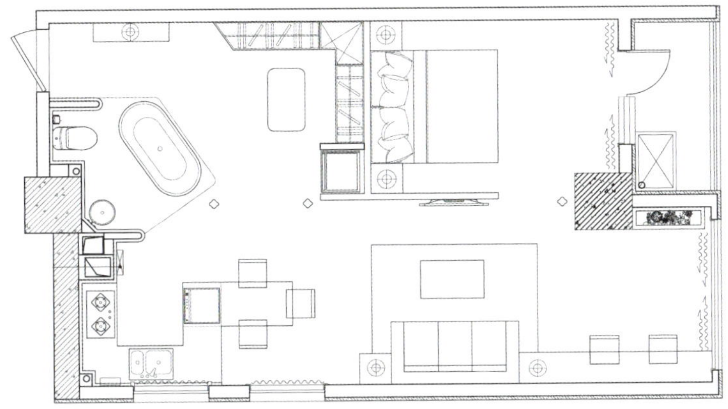

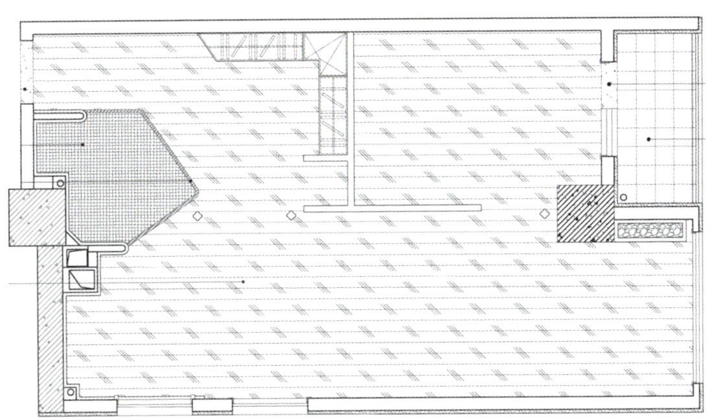

她是一个 35 岁的单身女性，在这个 50 平方米的空间里，浴缸终于不用委屈在狭小的私密空间里，而成为她居室的主体，无拘无束；盥洗室对面是她最爱的衣橱，她心爱的“生活艺术品”不用藏在封闭的木柜子里，它和她一样可以尽情释放自己的光芒；她不用经常下厨，偶尔心血来潮，奖励自己一顿美餐，“做”和“吃”也不用那么规矩，分的那么“标致”，完全可以在一起，不用装模作样……

起居室和用餐区不分彼此，她可以躺在沙发上享受美食，也可以坐在餐桌前看肥皂剧；书房在光线最好的地方与起居室相连，也可独立成一个空间，满足突然想认真写点什么的愿望；书桌对面是她的卧室，相对私密一些，她睡觉的时候需要一些安全感。

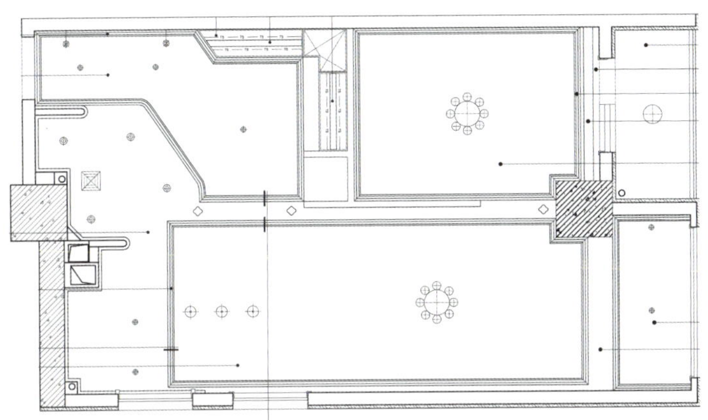

设计公司：台北拾雅客空间设计 设计师：许炜杰 项目地点：中国台湾 项目面积：75 平方米 主要材料：文化石、玻璃砖、烤漆玻璃、不锈钢、木皮、订制磁砖

MODERN FAIRY TALE IN WHITE

白色里的现代童话

The case was a house which is twenty-five years old, house height is low and is provided with beam barrier, thereby leading to poor initial conditions. In the premise conditions, the designer uses the arc wrapped beam techniques to resolve, and uses the mirror of a large area to expand the space visibility, thereby avoiding the unfavorable factors brought by beam cutting of the space. In addition, in order to increase the use area of public area, the designer further turns the original bathroom space for 90 degrees, thereby connecting with the private space in series to form a vertical -oriented neat feeling.

The male owner with cleanness bias expects clean, neat and simple home style, the designer creates multi-layered white space with material and element conversion under main shaft, and intimately takes into account the owner cleanliness, and the lacquer should be wrestled in the main entrance to improve the ease of maintenance.

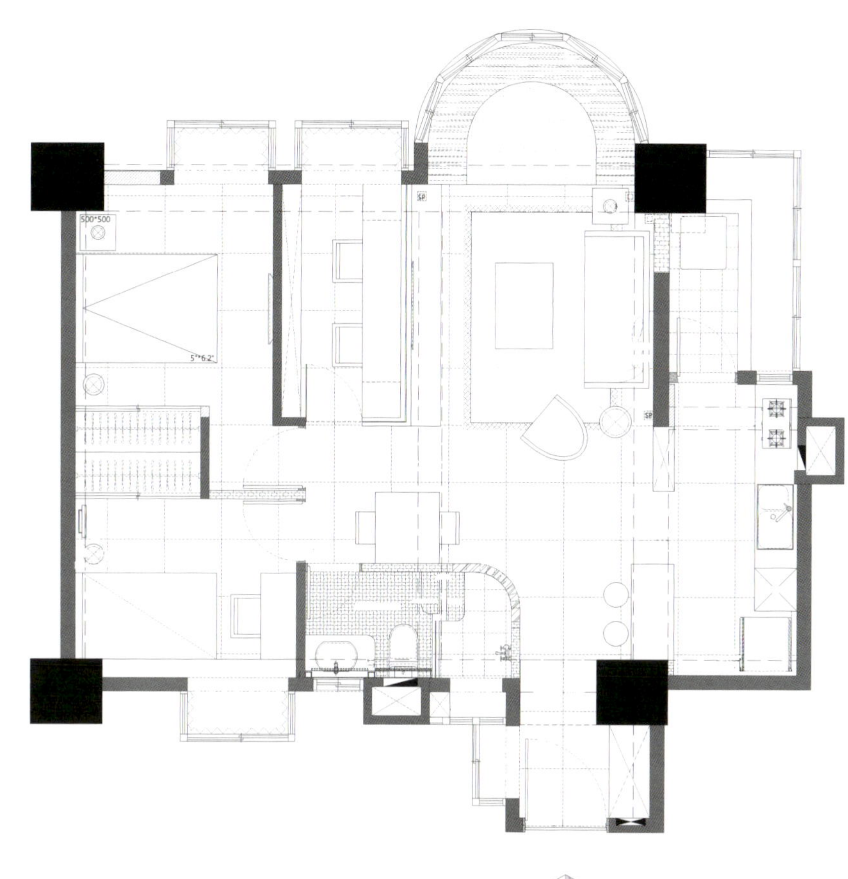

本案是一个二十五年的老屋，屋高低矮而且又有横梁的阻隔导致原始条件不佳。在这样的前提条件下，设计师用圆弧包梁手法来化解，同时用大面积镜面来扩大空间视觉，以回避空间被大梁切割所带来的不利因素。另外为加大公共区域的使用面积，设计师进一步将原有的卫浴空间九十度转向，与私密区域串联出垂直面向的利落感。

爱干净的男主人期待居家风格是净白、简单，设计师在此主轴下以材质、元素转换创造多层次的白色空间，且贴心考虑到屋主未来的清洁度，在主要进出区域斟酌漆料提升维护的容易性。

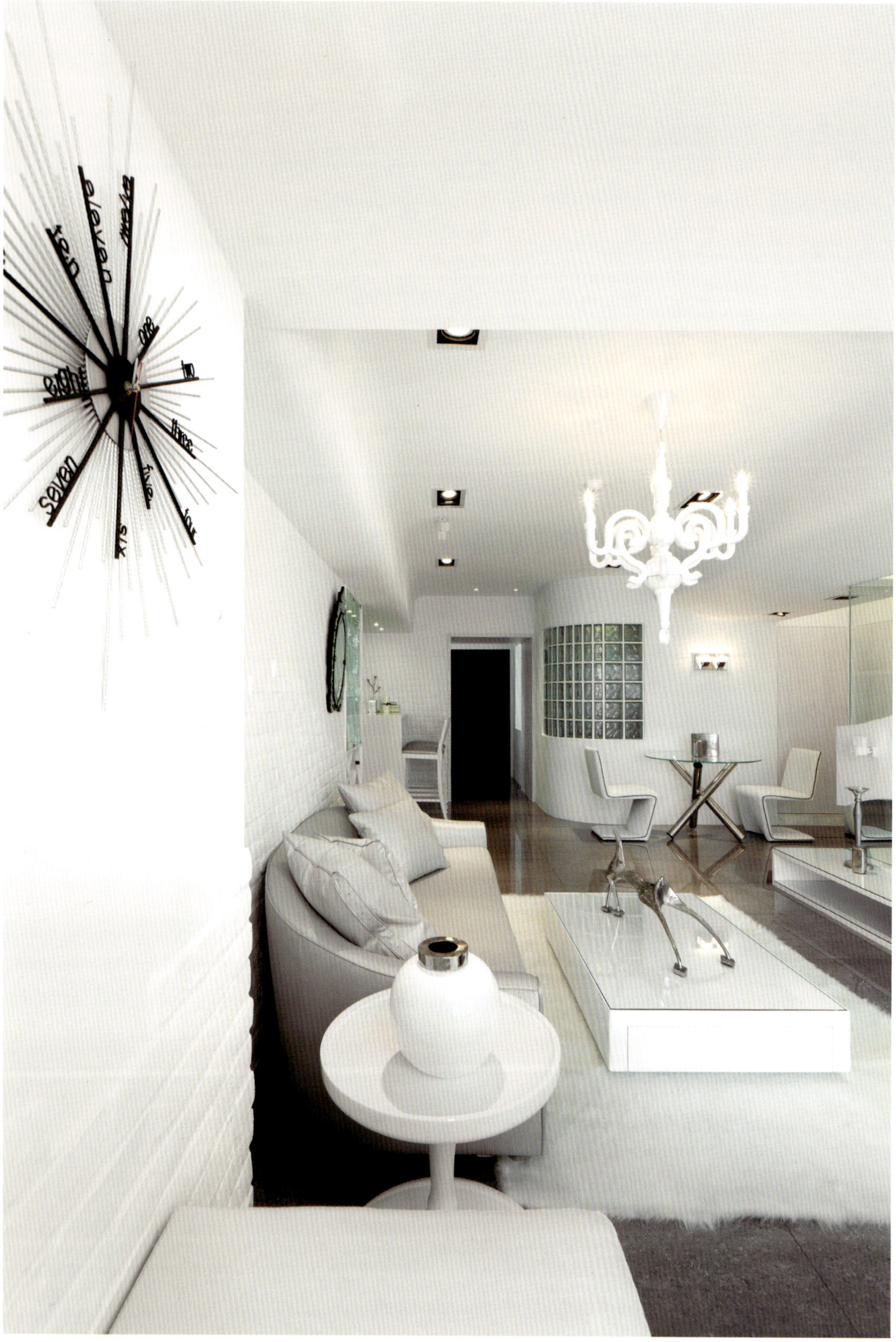
twelve
eleven
ten
one
two
three
four
five
seven

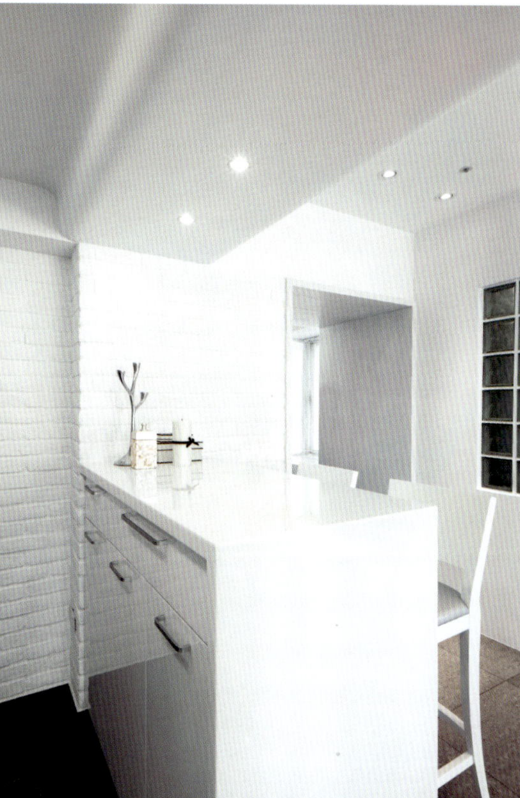

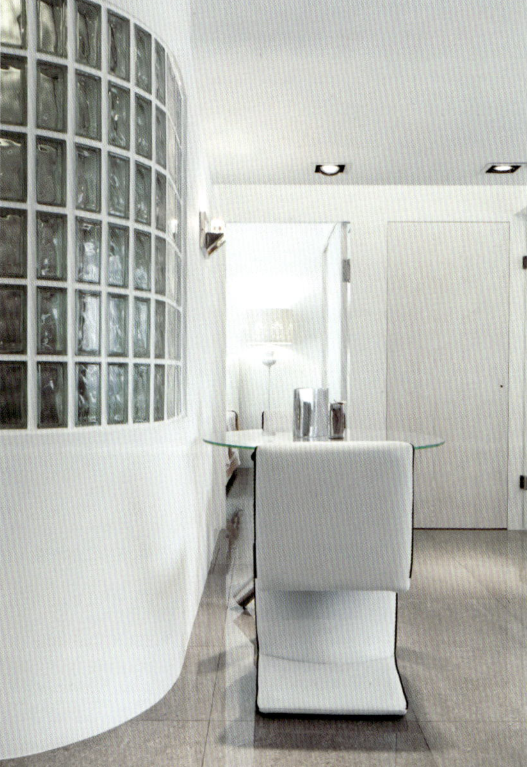

设计公司：广州市韦格斯杨设计有限公司 | 设计师：区伟勤 | 项目地点：中国山东 | 项目面积：70 平方米 | 主要材料：石膏模板、镜面不锈钢、马赛克、超白镜、透光贴膜

ZHONGHAI · INTERNATIONAL COMMUNITY MODEL ROOM UNIT A

中海 · 国际社区样板间 A1 户型

The model room belongs to Zhonghai Real Estate and is located in Shinan District of Qingdao. Qingdao has very strong colonial culture and is melted with modern and bustling atmosphere, thereby making it among the forefront of Chinese cities.

Model room of unit is affected by Qingdao geographical culture which gives birth to the modern interpretation techniques, it depends on Zhonghai • International Community Sales Department and provides owners with ornamental-based property.

It has a bright and spacious living room and star river stream type ceiling goes across. The kitchen and dining rooms are connected to create kitchen as main breadth and dining room as back breadth; it is tightly connected with the corridor. The master bedroom atmosphere particularly lies in its dynamic backdrop. The walls are provided with silver mirror stainless steel, and small space has general style. Soft decoration of kid room is matched rightly, which is really a crowning touch.

The entire space is based on white tone added with silver style. Dynamic modeling and perseverant innovation are the fundamental design of this case.

本样板间隶属中海地产，位于青岛市市南区。青岛极具浓厚的殖民地文化，融入现代繁华气息，使其跻身于中国城市的前列。

A1 户型样板间受青岛地域文化影响，使其诞生了现代的诠释手法，它正是依赖中海·国际社区售楼部而寄生，给业主提供观赏性为主的物业。

其拥有明亮宽敞的客厅，星河流带式的天花一划而过。厨房与餐厅相连，营造厨房主幅为餐厅背幅，且与走廊紧紧相连。主人房的气息在于其特别动感的背景墙。墙身加以银色镜面不锈钢，小空间有大气派。小孩房软装配合得恰到好处，实为点睛之笔。

整个空间以白色调为主，加以银色的格调。造型动感，锲而不舍的创新是本案设计的根本。

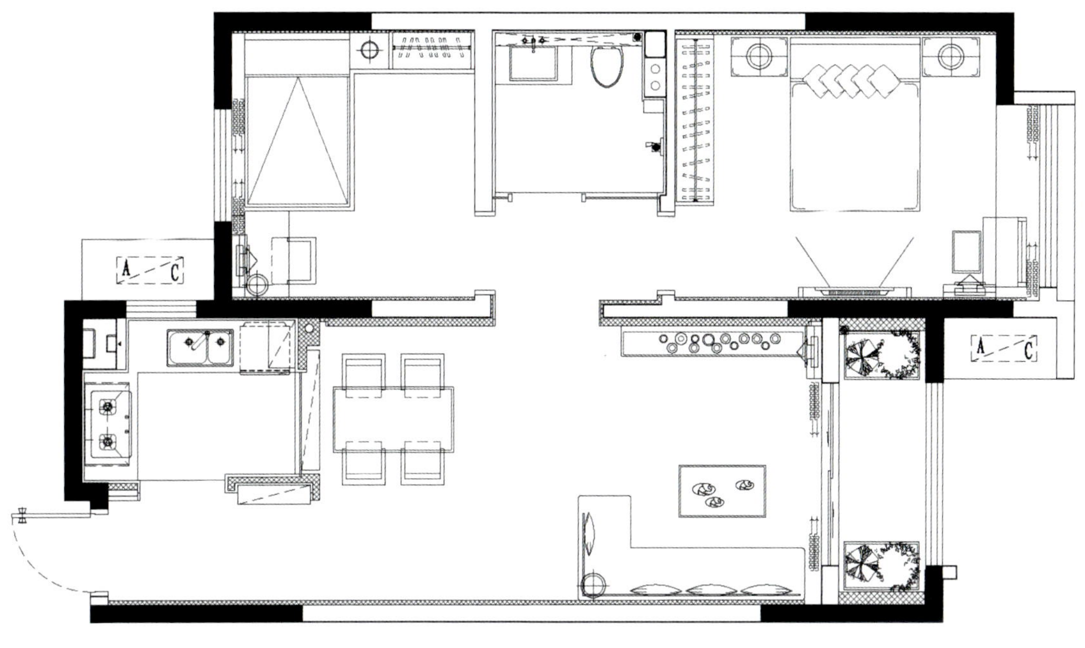
A C
A C

设计公司：东莞市形上几何室内设计有限公司 设计师：卢迅 项目地点：中国广东 项目面积：76 平方米

主要材料：米黄抛光砖、白色乳胶漆、深色柚木饰面板、柚木实木地板、米黄色墙纸、浅啡色墙纸

FENGTAI CITY CHINESE STYLE

丰泰城中式

House decoration style is sometimes not simply just a symbol; sometimes it also represents an attitude. The case belongs to a restrained and calm Chinese space, and is designed for middle - aged couple who have successful career and make idle at home. What they want and how to look at things tend to be more indifferent as the psychological journey of life in middle age, they wan to be quiet and seem to have settled down inner feelings, and their requirements on home can be closer to "scholarly temperament".

This case shows such scholarly temperament, dark teak furniture is cooperated with beige and brown cotton and imitation silk arts, and classical Chinese ink painting is used as background to blend a deep-restrained scene. The log jewelry, ceramics, ears and high simulation flower are scattered in obscure corners, however they create the effect of seeing unexpectedly, which is memorable with infinitude aftertaste.

家装风格有时候并不单单只是一种象征，有时候它也代表着一种态度。本案是一个内敛沉稳的中式空间，以事业有成、赋闲在家的中年夫妇为设定而做出的设计。如同人生的心理旅程，人到中年，想要的东西和看待事情的眼光都会更趋于淡泊，想要沉静一些，就仿佛已经沉淀下来的内心情感一样，对家的要求便会更接近“书香气质”。

本案所表现的就是这样一种书香气质，深色柚木家具搭配米白、咖啡色系的棉麻和仿丝布艺，以中国经典的水墨国画为背景，晕染出一幅深隽内敛的场景。那些原木的饰品、陶瓷、穗子、高仿真的插花散落在不起眼的角落，却营造出惊鸿一瞥的效果，令人难忘，让人回味无穷。

设计公司：北京圳鑫想向装饰设计有限公司 | 设计师：向东姝 | 项目地点：中国陕西 | 项目面积：80 平方米 | 主要材料：石材、玻化砖、壁纸

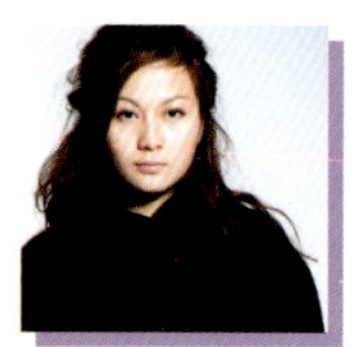

NEO CHINESE STYLE WITH LIGHT COLOR

淡彩新中式

Neo Chinese style does not purely refers to piling elements, it combines modern elements and traditional elements through the understanding of traditional culture, and creates things by rich traditional flavor with aesthetic needs of the modern. The case fully demonstrates this point and shows the charm and homeowner's taste.

In color, the designer selects elegant peacock blue and soft beige to decorate rose pink, thereby pointing out the theme of intrinsic elegance. The homeowner wants to decorate "home" into a portrayal of own life and symbol of artistic taste. Therefore, the case reveals not only just style change and fashion element, but also meaning of cultural accomplishment.

The designer rationally combines Chinese elements and modern lacquer materials to show culture details of Neo Chinese Style, which is restrained and thought-provoking. The space is generally elegant, calm and dignified with unique modern romantic spirit, tradition and modernity are naturally blended without any trace of affectation sense, and thereby people feel more comfortable.

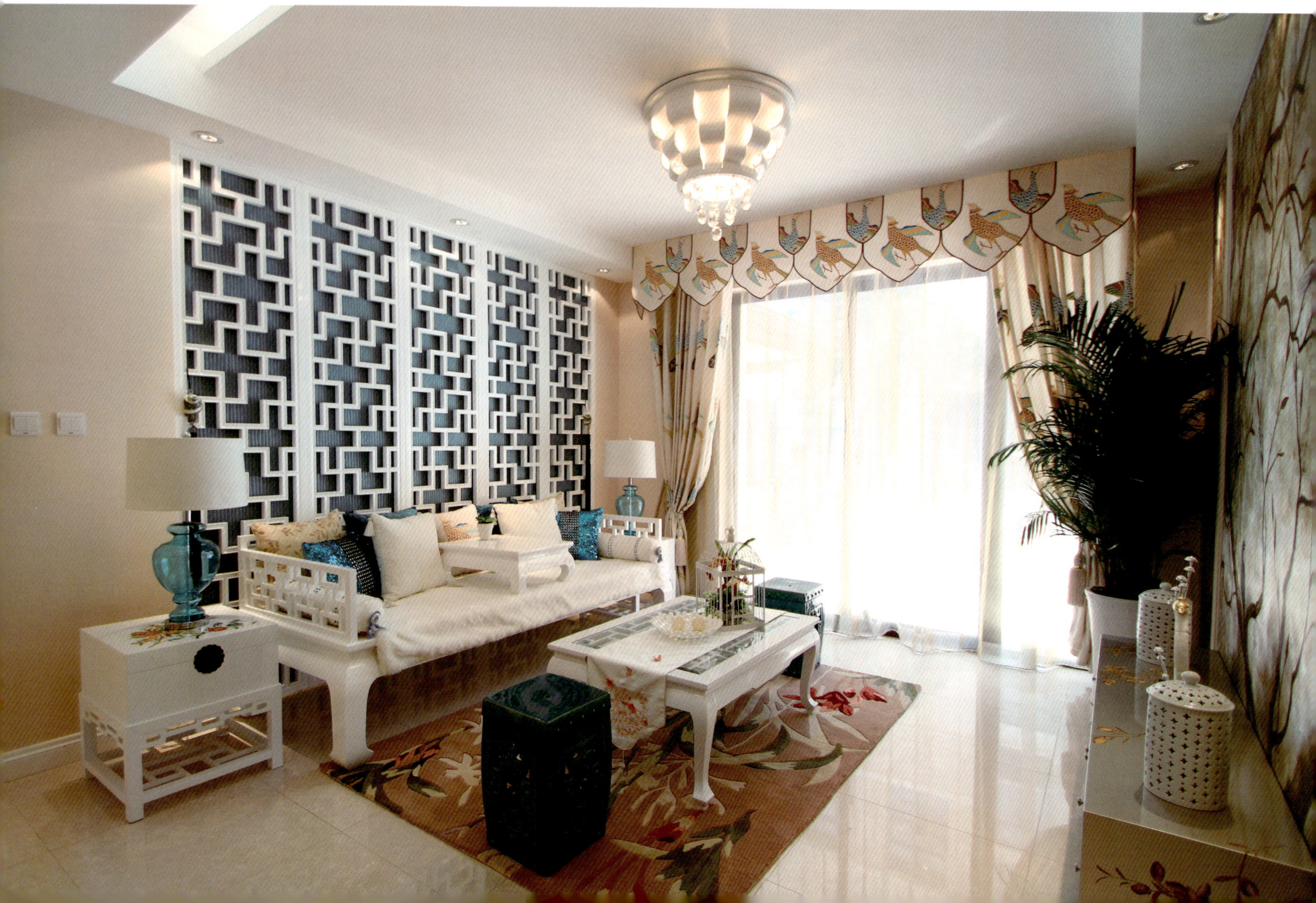

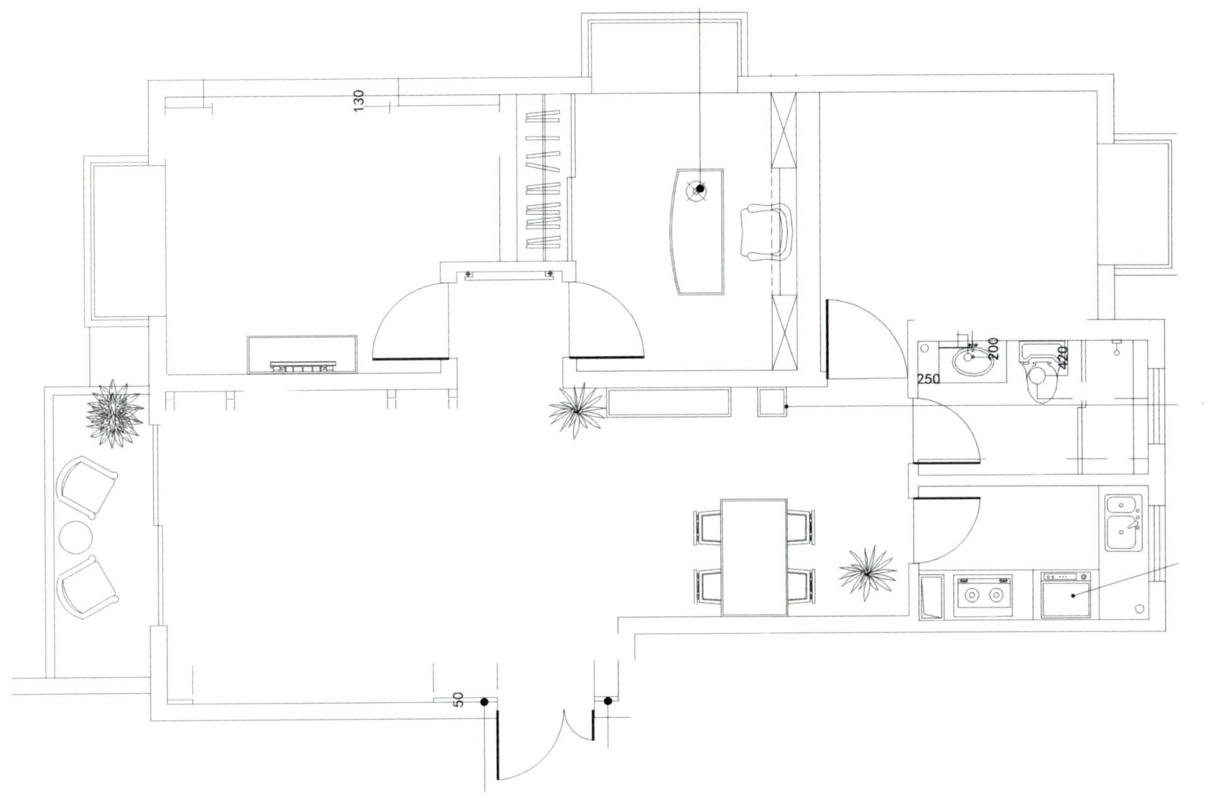
130
200
420
250
50

“新中式”风格不是纯粹的元素堆砌，而是通过对传统文化的认识，将现代元素和传统元素结合在一起，以现代人的审美需求来打造富有传统韵味的事物。本案充分体现了这一点，展现出新中式的魅力和房主的品位。

设计师在色彩上选择以清雅的孔雀蓝与柔和的米色来点缀玫粉色，点出内蕴风华的主题。房主希望将“家”打造成自己生活的写照和艺术品位的象征。因此本案流露出的不仅仅是一种风格变化和一个时尚元素，更是一种文化修养的内涵。

设计师将中式元素和现代漆艺材料合理搭配，表现出新中式的文化底蕴，内敛而耐人寻味。空间整体温文雅致、沉稳端庄，别具现代浪漫主义精神，传统与现代自然而然地融为一体，没有一丝矫揉造作之感，让人倍感舒畅。

Arletta
Arletta
Arletta

设计公司：东莞市形上几何室内设计有限公司 | 设计师：卢迅 | 项目地点：中国广东 | 项目面积：83 平方米 | 主要材料：米黄色抛光砖、胡桃木地板、浅色柚木饰面

EAST SEA CASTLE CHINESE STYLE

东海城堡中式

This is a graceful and elegant Chinese-style space, and elements with rich traditional Chinese complex are penetrated throughout the space, which embodies the space features while reflecting homeowner's taste and pursuit. Simple and smooth lines are used to outline the space from the public space of living room, restaurant to private space- bedrooms, the decoration delivers low-key, gentle and restrained temperaments with the original color of the materials, furniture and decoration of Chinese characteristics are arranged, thereby the house looks like a scene traced out by famous painter, which is both dynamitic and quite.

The choice of materials is wood -based, light-colored teak furniture is combined with beige and brown cloth art to enrich the space expression. Cloth fabrics made of different materials express different emotions with different textures, the whole space shows elegant and timeless qualities, and taste of the space and the pursuit of homeowners are completely exhibited through using Chinese traditional lamps combining wood and fabric, ink painting, ceramics, natural stone, ears and other ornaments rich in Chinese features as decoration.

这是一个娟秀典雅的中式空间，以富有传统中式情结的元素贯穿空间的始终，在体现空间特色的同时也反应房主的品位和追求。从公共空间的客厅、餐厅到私人空间——卧室，均以简洁流畅的线条来勾勒空间，以材料本身的色泽传达低调、温润、内敛的气质，再布置中式特色的家具和装饰，宛如名家手中描绘出的场景般，动静相宜。

在材料的选择上，以木料为主，浅色柚木家具结合米色、咖啡色系的布艺来丰富空间表情。不同材质的布艺织品以不同的质感表达着不同的情绪，加上中国传统式的实木与布艺结合的灯具，以及水墨国画、陶瓷、天然石、穗子等极富中式特色饰品的点缀，整个空间洋溢出典雅、隽永的气质，空间的品位和房主的追求尽显无遗。

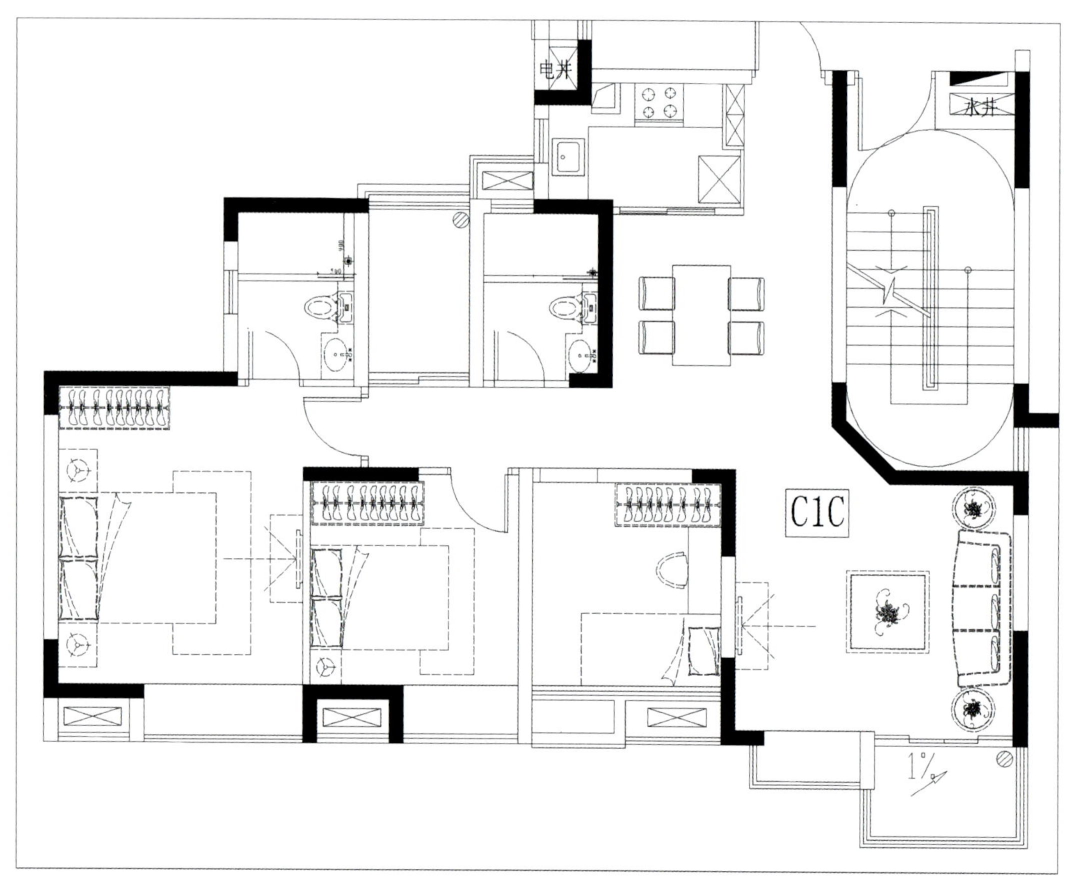
电井
水井
C1C
1%

设计公司：南京传古设计 设计师：于园 项目面积：92 平方米 主要材料：费昂纳墙纸、进口面砖等

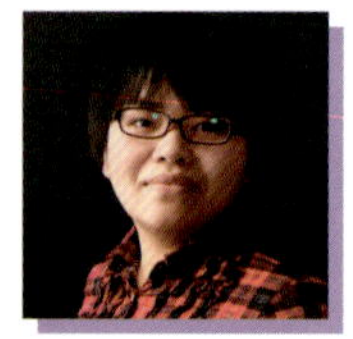

CHILDHOOD SWEETHEARTS–QUIET AND TRANSPARENT RAIN FOREST

青梅竹马·静透雨林

Some one says: in all those years, vegetation and woods are as many as forest, peaceful life is enjoyable.

She is provided with the fresh of dew in the forest; he is provided with rainforest and ancient wood handsomeness. They come from tropical rain forest; they have grateful monasteries of nature due to growth environment in childhood, and the words and deeds of their parent. The design focuses on green low-carbon environmental imperatively.

In the aspect of use functions, the designer starts from the low-carbon point of view for reasonable space partitioning and reduction of energy loss. The designer uses the tension design techniques, and fuzzes the borders of the space with unified white, thereby achieving the purpose of extended space.

In the aspect of visual communication, the broad framework of the entire space adopts pure white. White is pure and innocent, white is empty and represents of a life attitude of not being happy for obtaining, and not being sorrowful due to loss, thereby pursuing light, quietness, transparency and none tainting of common dust. The local part is matched with green similar to plant, a little rosy color and other bright colors to activate the space. Green shows his calm and generous feature as the green lawn, and she is cranberry dotted in the lawn, and they depend on each other. All methods are hollow except cause and effect, being kind to nature is being kind to us.

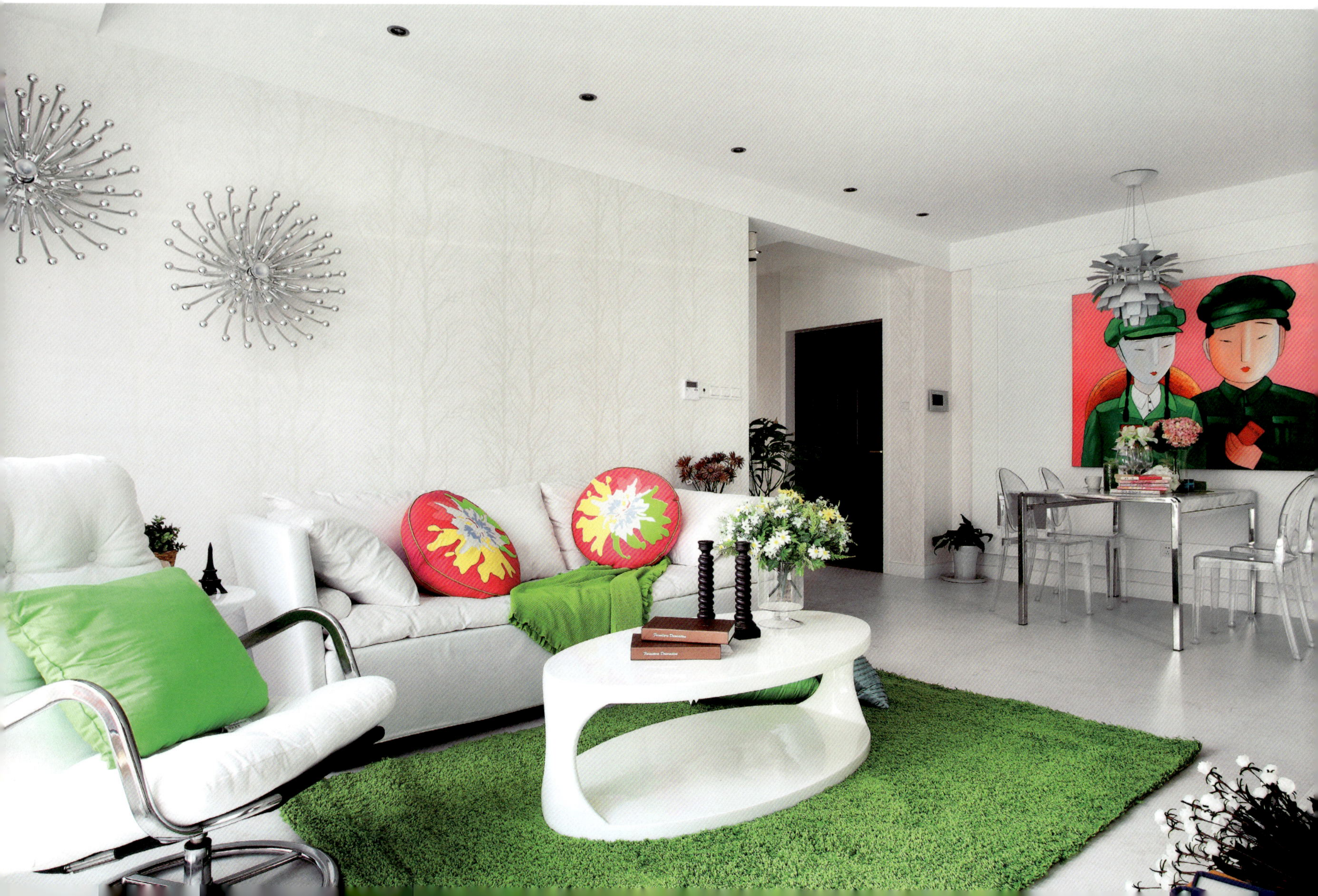

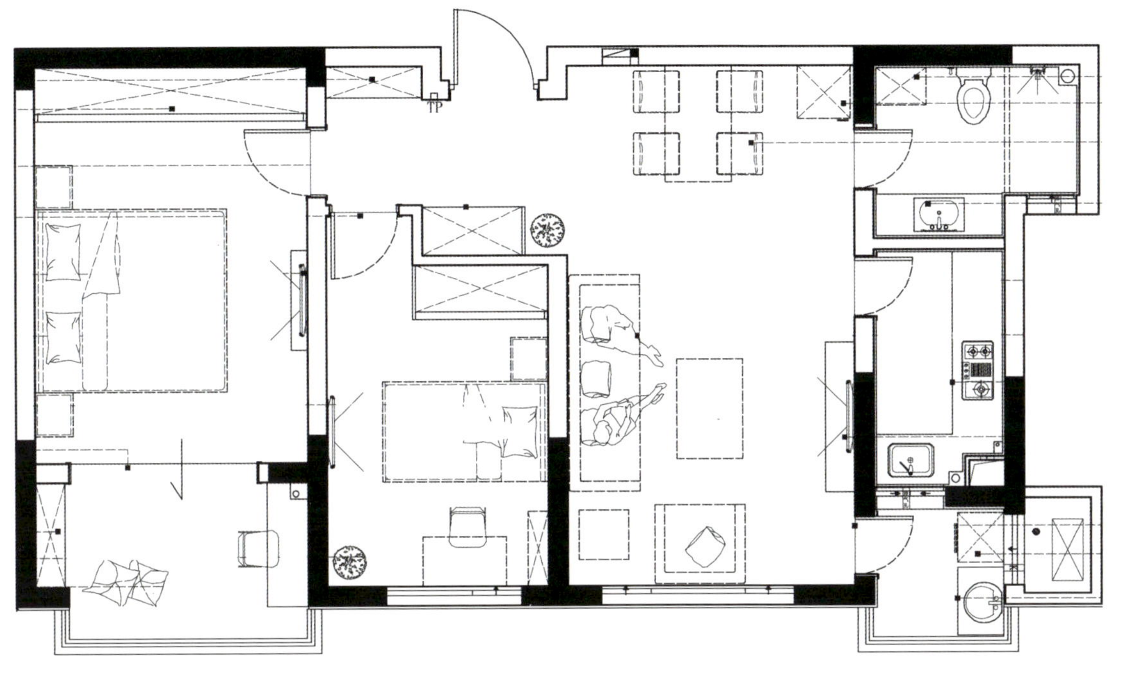

有人说：那些年，草木森然，岁月静好。

她，带着林间露水的清新；他，有如雨林古木般英挺。他们来自热带雨林，打小的生长环境，父母的言传身教，他们耳濡目染对大自然心存感激。此次设计，绿色低碳环保势在必行。

在使用功能方面，从低碳角度出发，空间分割合理，减低能源的损耗。设计师采用极有张力的设计手法，用统一的白色模糊空间的边界，从而达到延伸空间的目的。

视觉传达方面，整个空间大框架使用纯白色。白色纯净无邪，白色是空无，是得之不喜，失之不悲的一种生活态度，追求光明、静透、不染凡尘之垢。局部配以类似植物体的绿色及少许玫红等鲜亮色彩色活跃空间。绿色是他的沉稳与宽厚一如那青软的草地，而她是点缀其间的小红莓，互相依偎。万法皆空，唯有因果不空，善待大自然，即是善待自己。

设计公司：朗昇空间设计 | 设计团队：袁静、钟福建等 | 项目地点：中国广东 | 项目面积：68 平方米 | 主要材料：大理石、玻璃、地砖、镜钢等

ZHONGSHAN HELEN MODERN STYLE MODEL UNIT

中山海伦堡现代风格样板房

Small as this project area is, there is a structure with double balconies, double bedrooms, a bathroom and a kitchen, with all functions, full but not crowded.

Because of the small area and many functions, the designer makes best use of the advantages and bypasses the disadvantages, and digs up the space to plan the layout reasonably, to meet the life requirement. A modern technique is adopted in this project. The designer attaches importance to the material quality, the workmanship, as well as wonderful and rich adornment details, to express a modern and splendid beauty in the small space.

The porch, the guest room, and the dining room space are skillfully connected by large marble, steel mirror, crystal chandeliers and other objects, to make the space wholly present to be simple and fashionable. The clean and clear feature, matched with decorative details full of life, makes strong life breath come to people's face.

People entering the room, a large area of black steel mirror in the public space, the exquisite crystal chandelier and wooden grain board materials, and the fabric sofa and seats are perfectly matched with each other, building a fashionable and elegant, concise and bright life atmosphere, and making people have a modern and splendid beauty in this small space.

本案建筑面积虽小，但却有双阳台、双卧、一卫一厨的格局，各项功能俱全，丰满且不拥挤。

因为面积小而功能多，于是设计师扬长避短，挖掘空间来合理规划布局，以满足生活需要。本案采用现代风格手法，注重物料品质、做工，以及丰富精彩的配饰细节，以表现时尚小空间中的现代奢华之美。

玄关及客、餐厅空间通过大块面的大理石、镜钢、水晶灯等材质造型有机地联系在一起，使空间整体上呈现出简洁时尚、干净利落的特色，再配以极具生活化的装饰细节，浓浓的生活气息扑面而来。

入户后，公共空间里大面积的黑色镜钢、精美的水晶吊灯与木纹板材、布艺沙发和座椅完美地搭配在一起，营造出时尚优雅、简洁明快的生活氛围，让人感受小空间的现代奢华之美。

设计公司：武汉多维空间艺术设计有限公司 | 设计师：张晓莹、陶清明、熊丹 | 项目地点：中国山西 | 项目面积：56 平方米

主要材料：柠檬黄清玻、玫瑰金镜钢、巧克力色漆、超白砖、白色木作

TAIYUAN TIANSEN PEACE NUMBER ONE MODEL UNIT M DWELLING-SIZE APARTMENTS

太原田森安宁一号样板房 M 户型

This project area is not large, and potential customers are a couple born in 1980s. So the designer defines the space as "chocolate theme", according to the preference and requirements of the potential customers. The space color tone is expressed by chocolate color plus lemon yellow plus cherry red, to express the newly-married romantic appealing and beautiful emotion of life. At the same time, the designer makes a bit greater change for the space arrangement, and builds a retractable storage bed to meet the functional requirement of life.

Placement of the space top is the biggest highlight of this model unit. The designer manages the white and coffee color to weaken the height of beams, and weakens heavy beams through the chandelier shape of "white chocolate wafers", to add theme fun to the space. The sofa background is made of chocolate squares shape blended by coffee color and milk white, and fully reveals the household affection and comfort, which makes the host take off a day's exhaustion once coming back home and come to the beautiful family life.

餐厅
客厅
卫生间
厨房
卧室
旋转电视

天花平面布置图 S=1:40

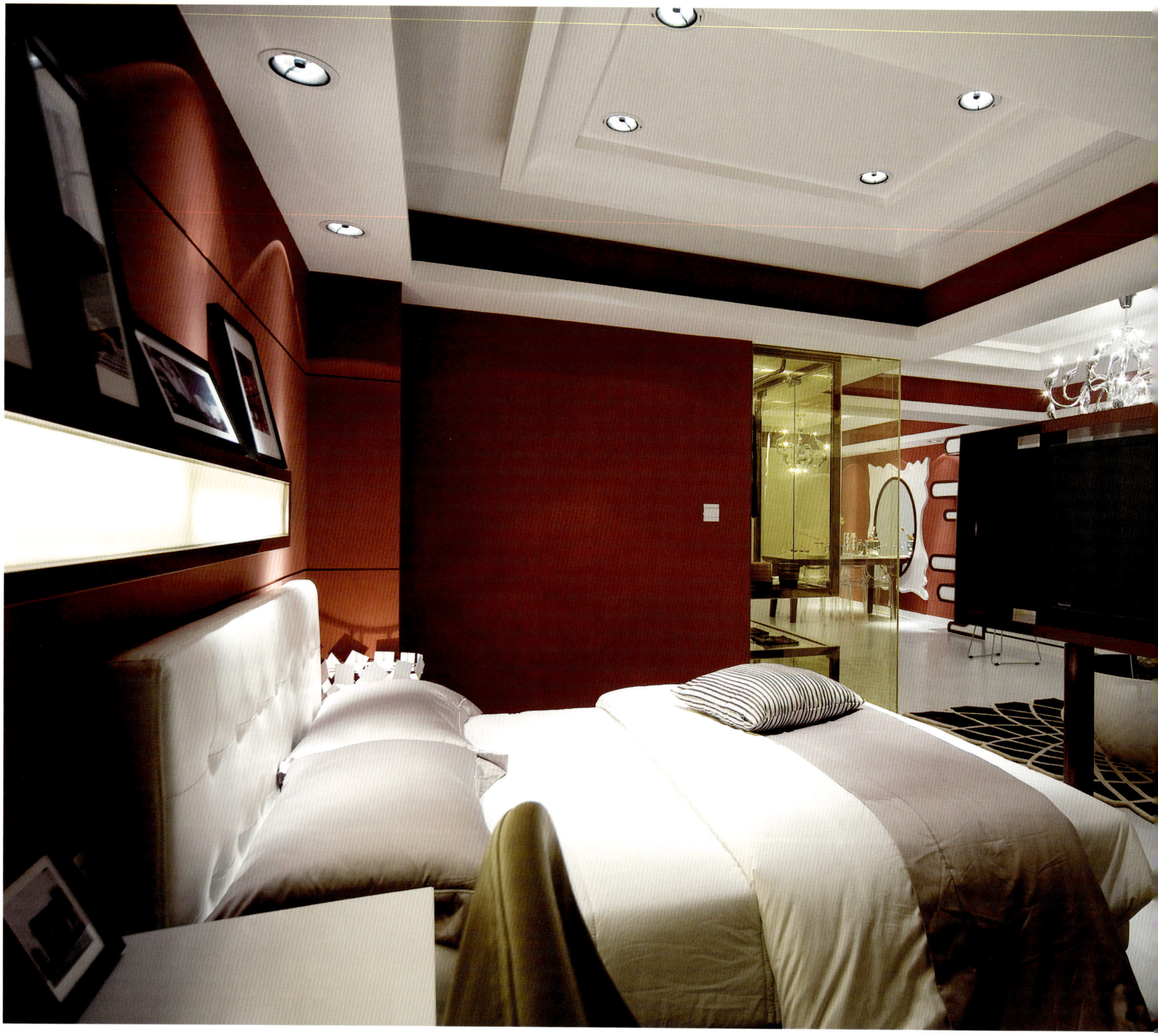

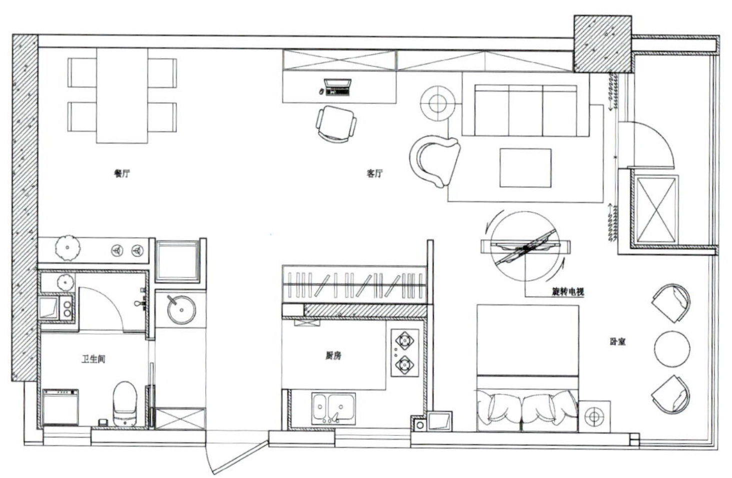

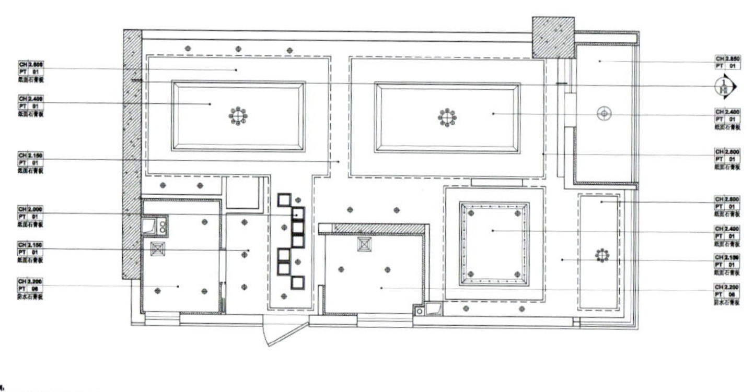

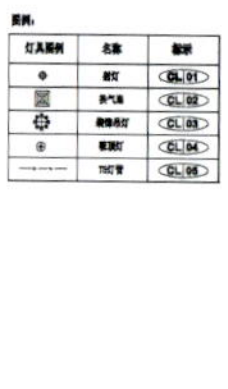

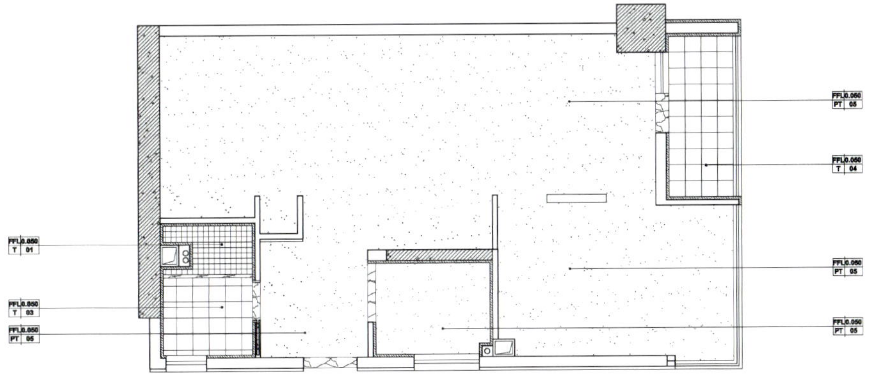

地面布置图 S=1:40

该户型面积不大，虚拟客户人群为一对 80 后新婚夫妇。设计师根据虚拟客户的人气喜好和需求，将空间定位为“巧克力主题”。空间基调以巧克力色 + 柠檬黄 + 樱桃红来体现，表达新婚的浪漫情趣与生活的美好情感。同时，设计师对空间布局做了较大的改动，并做了一个可收放的收纳客床来满足生活上的功能需求。

空间顶部处理是此样板房的最大亮点之一，设计师利用白色和咖啡色的分色处理，弱化了梁体的高度，然后用“白色巧克力薄饼”的造型吊顶弱化梁体的厚重感，为空间增添了主题情趣；沙发背景以巧克力方块造型加之咖啡色与牛奶白调和，充分地体现出居家的浓情、惬意，使主人一回到家里就甩脱一天的劳累感，融入到美味的家庭生活中来。

设计公司：武汉支点环境艺术设计有限公司 设计师：杨坤、段威 项目地点：中国湖北 项目面积：47 平方米

GOLDEN HARBOR

金色港湾

This project is defined as modern style, where the space of 47 square meters could be built to 90 square meters. The skillful combination of black, white and grey goes through the whole space wall, furniture and flooring, making the whole space have a great impact on vision and fashion.

The designer unites the dining room and the living room, maximizes the vision on condition that all functions' requirement is met. The design without partitions, plus the application of metal, mirror and other materials, not only extend the space but also optimize the space structure. White is the major color, adorned by black and grey, which adds more of being cool and fashionable. In order to avoid the boring feeling, the sofa setting wall is decorated by stripy wallpaper with khaki color, not abrupt, but shows the artistic temperament of the decorative fresco. The space beside the living room is made into a semi-open book room, white shelf, bookshelf, and black seats are not only the furniture, but also the ornaments.

The design of the bedroom is a geometric art space. At first there is a dressing table right in front of the bed, and the white partition of irregular shape acts in cooperation with geometric metal tea table by the bed, which is fashionable and personalized and adds a tinge of elegance to the space.

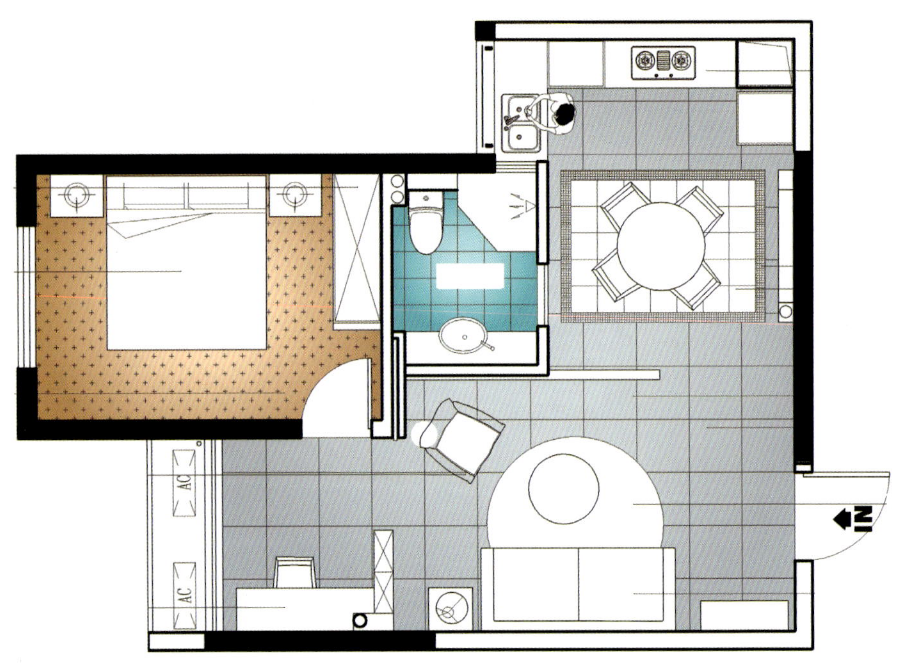

本案以现代风格定义空间，在 47 平方米的空间里打造 90 平方米的空间体验，利用黑、白、灰的巧妙搭配贯穿整个空间的墙面、家具和地面，使整个空间极具视觉冲击力和时尚感。

设计师将餐厅和客厅空间一体化，在满足功能需求的前提下让视觉达到最大化，无区隔的设计加上金属、镜面等材料的运用，在延伸空间感的同时也优化了空间格局。白色为主，黑、灰色为点缀的搭配，更添几分时尚酷感。为了避免单调，沙发墙采用卡其色系的条纹墙纸装饰，既不会突兀，反衬出装饰壁画的艺术气质。客厅一旁的空间被用来做成一个半开放式的书房，白色的搁板、书架和黑色的座椅既是家具，也不失为艺术点缀。

卧室的设计是几何体的艺术空间，首先是正对床头的妆台设计，采用不规则形状的白色隔板呼应床头几何形体的金属案几，既时尚又显个性，更为空间增添几分优雅气质。

设计公司：深圳市昊泽空间设计有限公司 | 设计师：韩松 | 项目地点：中国广东 | 项目面积：88 平方米 | 主要材料：黑金砂、枫木地板、铁刀木饰面、墙纸

FU TONG TIAN YI BAY THE SECOND PHASE MODEL UNIT BUILDING 3 DWELLING-SIZE C

富通天邑湾二期样板房 3 栋 C 户型

The fashion of black, white, and grey builds a space with a quiet noble, a sense of quality, to show more of its taste. A clean and orderly white and pure gives people a comfortable feeling. Wooden color brings a natural fresh with thick natural flavor. Black furniture matches with the bookcase background in low-key colors, serene and comfortable. The designer builds a concise, bright and angular space, and expresses these comfortable features to the extreme.

Concise and distinct color plus concise and smooth lines build a trace of freshness and brightness, which brings urban people in a fast rhythmic life with feasting and revelry a comfortable feeling, and makes people think home is very sweet as well. This is a kind of mood, also a kind of atmosphere, as well as a pleasant feeling of home.

黑、白、灰的时尚，营造出的空间高贵中带着安静，有质感，更显品位。干净利落的白色，纯洁中总是给人舒服的感觉；木色带着大自然的清新，有着浓浓的天然味；黑色的家私配以低调色彩的书柜背景，沉稳而舒适。设计师以简单的设计风格，打造出一个简洁明朗、棱角分明的空间，并将舒适宜人的特色抒发得切实到位。

简明的色彩加上简约流畅的线条，营造出一份清爽、明朗，让生活在快节奏、灯红酒绿的都市人一种舒心的感觉，也让人更觉得家很温馨。这是一种情绪，也是一种氛围，让人轻松愉悦的家的感觉。

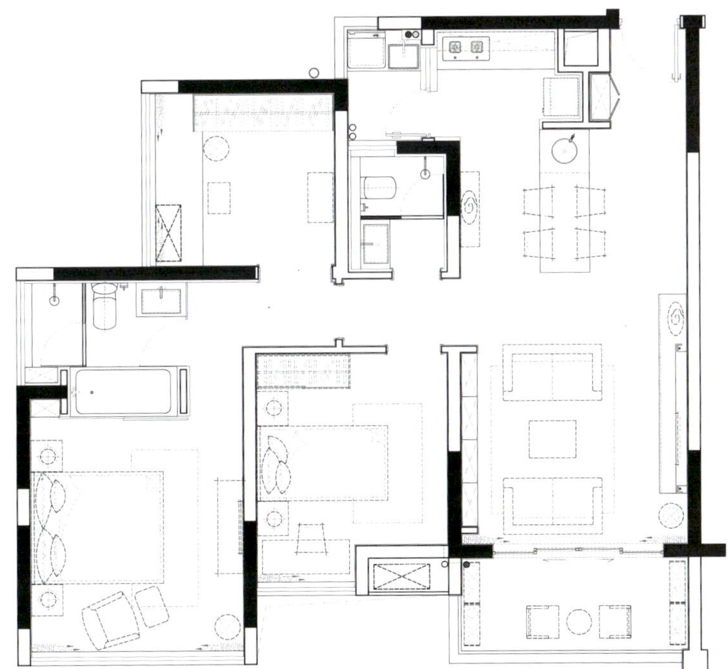

设计公司：春雨时尚空间设计 | 设计师：周建志 | 项目地点：中国台湾 | 项目面积：69.3 平方米 | 主要材料：银狐大理石、镜面、壁纸、镜面

DA ZHI LIANG MAO LIN RESIDENCE

大直良茂林公馆

The pattern of two-family-house usually cannot build the porch area due to the gate opening position limited, and there is a sense of dislocation on the vision, so the designer resolutely abandons all gates, carefully avoids the disadvantage of the gate facing the gate, slightly adjusts the major and minor bedroom doors, and designs the complete storage space.

In the sitting room, taking the house owner family fond of reading into consideration, the designer makes a special cabinet design behind the sofa wall, half of which is decorated by the curtain in order to avoid falling dust, the rest of which is reserved for storage space by transcurrent doors, in order to realize the household storage by a clean-cut technique.

Entering the restaurant, blue baking paint glasses and magnetic board make the colorful word, creates space of dual purpose with both the dining room and Internet space, while the original position of the door facing the door is transformed into a cabinet, in order to make the whole family have a center getting together. In addition, in order to eliminate the original structure of the beams, the designer makes a leaning and undulating ceiling, and meets the requirement of lighting in this area by embedded lights and indirect lighting.

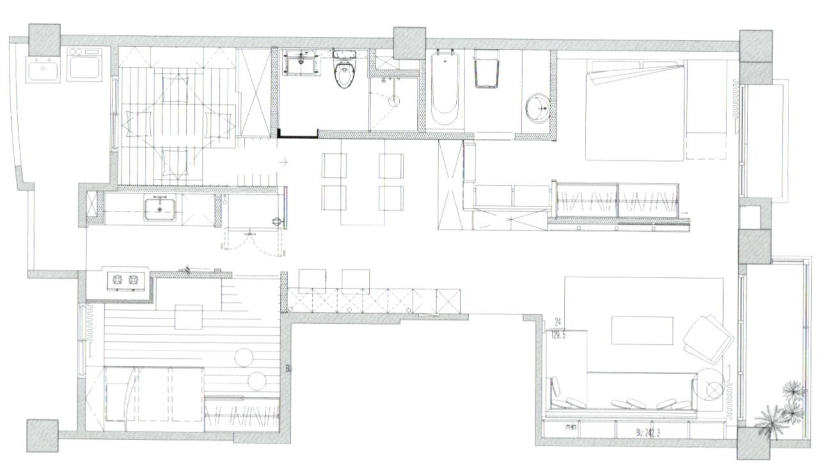

双拼格局因受限于开门位置，往往无法营造玄关场域，又会造成客、餐厅在视觉上的断层，设计师果断退掉所有的门，还细心地回避掉门对门的风水疑虑，并利用主卧门片的微调，设计出完整的储藏空间。

在客厅里，因考虑到业主一家人喜爱阅读的习惯，设计师在沙发背墙后施以特殊柜体设计，上半段辅以帘子遮掩、预防落尘，中、下半段则以横推门片预留置物容量，以利落手法实现居家收纳。

步入餐厅空间，以蓝色烤漆玻璃和磁板打造的缤纷，创造出餐厅与上网的两用空间，而原本与和室门对门的位置，则转换成展示柜体，让一家人有了凝聚的中心点，另外为了消弭原格局中的大梁，设计了倾斜起伏的天花，藉由嵌灯及间接光源满足此区的照明需求。

设计公司：之境室内设计事务所 设计师：廖志强、张静 项目地点：中国四川 项目面积：76 平方米 主要材料：乳胶漆、瓷砖、马赛克

ONLY ONE PEACH

蜜桃不二

The highlight of this project is to enlarge the space. The designer makes use of the curved wall to enlarge the originally narrow restaurant space, divides the bathroom into the dry area and the wet area, and places the beautiful basin outside. The table style is to act in cooperation with early-designed curved wall and to ease a sense of shortness cased by the passageway. The designer not only takes the space atmosphere into consideration but also makes a lot of work act in cooperation with the design, such as the decorative wine rack on the wall of the restaurant, the circular dressing mirror, the curved rack and so on.

The highlight is to how to make the door of bathroom. The limitation of the space forces the bathroom not be able to adopt the normal door, so the designer boldly adopts the double acting door, on both of which is pasted with the carefully selected poster to make it conform to the design theme and add space fun as well. The bedroom design keeps up the whole design style, which not only acts in cooperation with the curved design but also avoid some structural problems in architecture.

此案的设计亮点在于空间的扩大化处理，设计师利用弧形墙体，把原本狭小的餐厅空间扩大，另外将卫生间进行干湿分区，把漂亮的洗脸盆进行外置。餐桌的样式是为了呼应前期设计的弧形墙体，另一方面也是为了缓解过道所带来的急促感。设计师不仅考虑了空间的感受，在设计的呼应上也做了大量的工作，如：餐厅墙面的装饰酒架、圆形的梳妆镜、圆弧形的搁物架等。

最大的亮点是卫生间门的处理，空间的局限性迫使卫生间不能采用常规的门，设计师就大胆地采用推拉门，并且在推拉门的两面都贴上精心筛选过内容的海报，让它更贴合设计主题，同时也增添空间情趣。卧室的设计延续整体的设计风格，呼应弧形设计的同时，还巧妙地规避建筑上的一些结构问题。

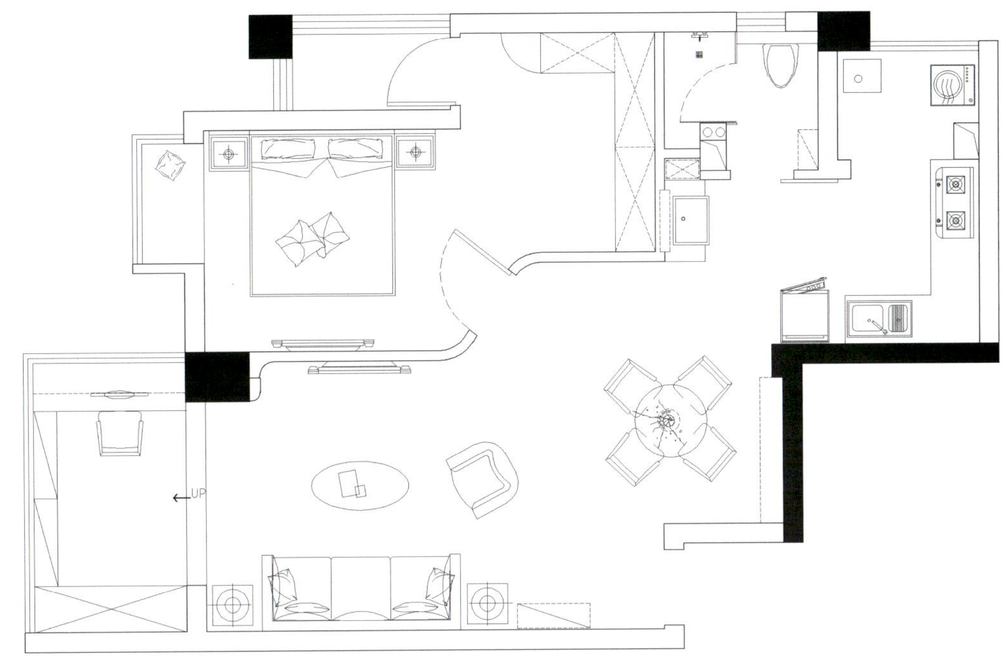

设计公司：尚扬理想家空间设计股份有限公司 | 设 计 师：陈元旻 | 项目地点：中国台湾 | 项目面积：69.3 平方米 | 主要材料：秋香木皮、黑檀木皮、抛光石英砖

NEI HU YOU RESIDENCE

内湖尤公馆

This project is located in highly intensive national residential area. Limited by the height of the building itself and narrow structure, the designer reduces and hides the traditional door as the design theme by building the penetrability of the public space, reconsiders each functional area and partition pattern, and constructs a small dwelling-size residential space with full functions and smooth moving lines.

In order to have PhD (ABD) sit in a space full of edification with a book in his hand, the designer renovates the 15-year-old house, and raises reading atmosphere like famous book stores after completing the basic project work. The sedate and solid double bookcase, going with the book room space for both storage and display, makes reading books become a natural music of life, where life is a pleasant life experience.

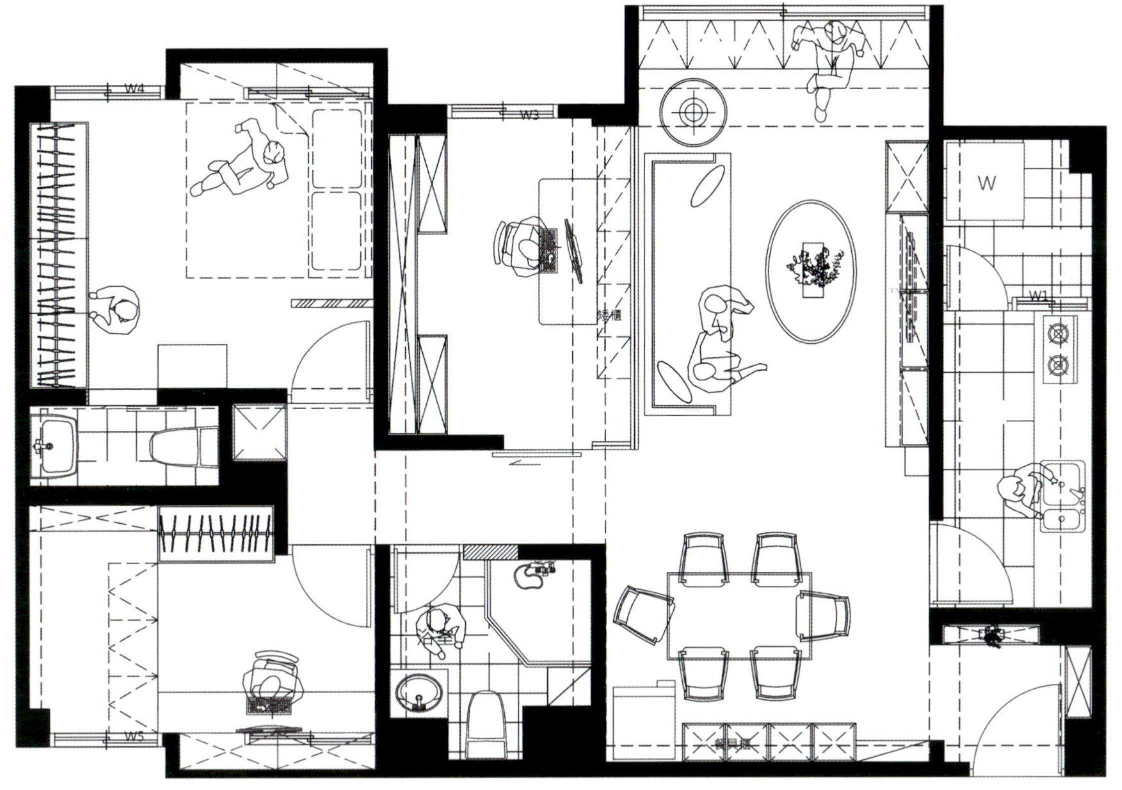
W4
W3
W
W1
W5

此案位于高度密集的国宅住宅区内，受限于建筑物本身的高度及狭小局促的格局，设计师以营造公共空间的穿透性及对传统门片进行减量与隐藏作为本案的设计主轴，重新思考各功能区的必要性与分割方式，创造出机能充足且动线流畅的小户型住宅空间。

为了让准博士生坐拥书香熏陶的质感空间，设计师将房龄十五年的老屋翻新，在基础工程完备后，导入知名书店的阅读氛围，以扎实且沉稳的双层书柜，搭配出收纳与展示兼具的书房空间，让读书变成生活的自然乐章，至于生活则是惬意的人生体验。

设计公司：春雨时尚空间设计 | 设计师：林郁卿 | 项目地点：中国台湾 | 项目面积：86 平方米 | 主要材料：茶镜、烤漆、人造皮、大理石、铁件屏风、壁纸、木皮

ZHONG HE NAN SHAN ROAD QING GONG RESIDENCE

中和南山路钦公馆

The tawny mirror and white baking paint with a wide range of length, adorned by a real sense of leather, the porch builds a rhythmic sense of moving by means of different materials and lines, and goes along the vertical bright surface, so the scene of a sweet small family appears. Coming to a pleasant corner getting the most lighting—under the windowsill, the host and hostess read books and magazines, and the children surf the Internet and play games, a relaxing family day, family's favorites.

The designer offers this not spacious living room an enlarged effect. The marble extending to both sides widens lines of the living room. In the pure white world, furnishings with a tinge of fashion goes with sweet wooden tea table and a warm visual effect caused by woods brings a harmonious emotion for this small family. Along the lines of the tawny mirror, the designer plans the moving lines getting to a dining ground and private area, where wooden tables and chairs and unique and exquisite lights brings content to this dining space, and acts in cooperation with the small living room, recording the family's sweet interaction.

不等宽的茶镜、白色烤漆，缀上富有质感的皮革，玄关以异材质与线条创造律动感，顺着垂直的亮面，浮现温馨小家庭的景象。走至一个汲取最多光亮的惬意角落——窗台下，男女主人翻阅书刊杂志、孩子上网、游戏，轻松的纯白家庭日，是全家人的最爱。

设计师以纯白赋予不够宽阔的客厅放大的效果，向两端延伸的大理石拉宽了客厅线条，放眼洁白的世界，略带时尚感的陈设搭配可爱的木制茶几，木质温暖的视觉效果点出小家庭的融洽情谊。沿着茶镜线条，规划出通往餐厨场域与私人场域的动线，木造桌椅与别具特色的精巧灯饰让用餐区有了主题，呼应着小客厅，一起记录家人之间的温馨互动。

reece

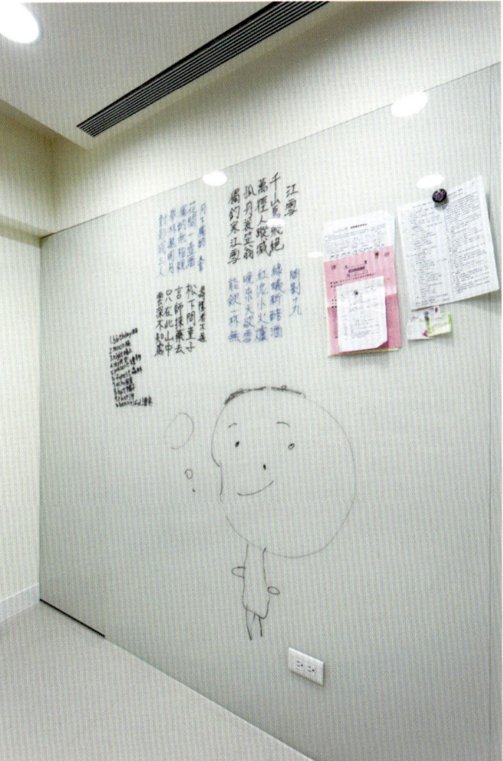

设计公司：绝享设计工程有限公司 | 设计师：绝享设计团队 | 项目地点：中国台湾 | 项目面积：53 平方米 | 主要材料：人造石、苏菲亚大理石、茶玻喷砂、美耐板

HUANG RESIDENCE ON WENQUAN ROAD

温泉路黄公馆

Creation of maximum function in the space and active utilization of space simplified lines for division will become the most important tasks of the designer since the space area is small. The designer displaced the kitchen originally located on the balcony, therefore the sense of space can be improved, the outside of the window is supplemented by plants, chandelier with plant vegetation also can be added on the table to bring a trace of green to space, it is matched with the wine cabinet of black mirror decoration on the right so that the entire space has more texture, a playful and relaxed atmosphere can be created, and it seems that even time stops in her tracks.

The overall space plans respective access routes, a double-sided compartment cabinet is arranged behind the sofa, and doors and cabinet adopt integrated planning to simplify the lines of the space. The compartment cabinet behind the sofa is also added with practical electrical cabinet on the side, indirect lighting is moderately marked to light up the entire space.

Green is abundant in white, a wonderful enjoyment of hot spring pool is available although the area is small, and it can be said that the room is small but perfectly formed.!

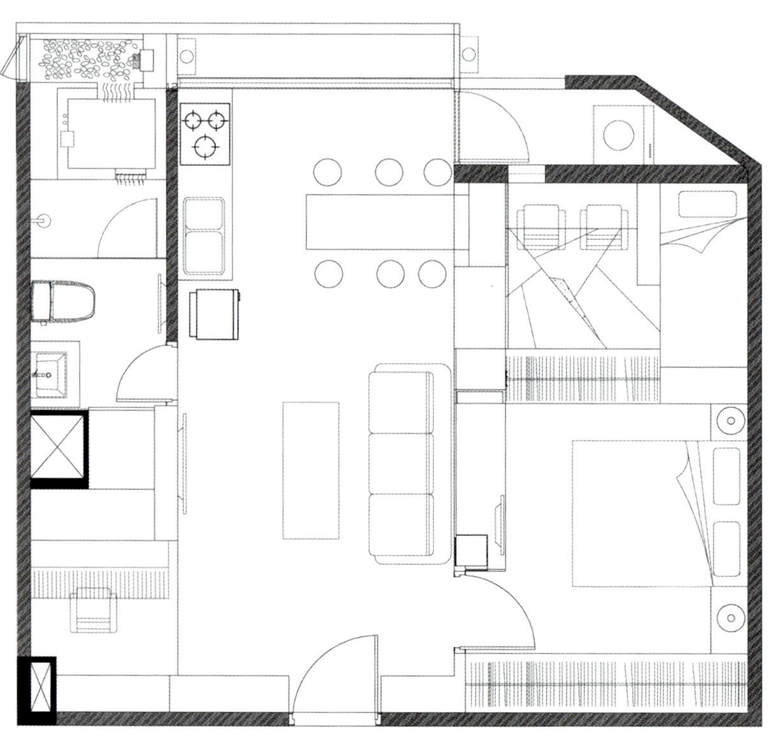

空间面积很小，因此，创造空间最大使用机能、活用空间简化线条分割便成为设计师的第一要务。设计师对原位于阳台的厨房做了移位处理，空间感也因此得以提升，窗外辅以绿植，餐桌上也以可加种植物的吊灯装饰，为空间带来丝丝绿意，搭配右侧以黑镜装饰的酒柜，让整个空间显得更有质感，同时营造出轻缓、放松的氛围，似乎连时间都停下了脚步。

整体空间规划出各自的出入动线，沙发后面是双面隔间柜，门和柜体采用一体式的规划，简化了空间的线条。沙发背后的隔间柜，也在侧边加入实用的电器柜，适度地打上间接照明，让整个空间都亮了起来。

在一片白色中，绿意盎然，面积虽小，却有着温泉汤池的绝妙享受，真正是：麻雀虽小，五脏俱全了！

设计公司：台北嵘特设计 | 设计师：王绍羽 | 项目地点：中国台湾 | 项目面积：66 平方米 | 主要材料：抛光砖、白烤漆、灰玻璃、美耐板、皮革绷布、木作、镜面

RETURNING SPACE TO CONCISE ORIGIN

让空间回归简洁的原点

In this case, the designer makes design closer to the lifestyle, there is no exaggerated and complicated decoration, the space is simplified from complex, white was selected as the main base color, style expression with unique charismatic charm can be carved out by the aid of neat lines of personalized furniture, the unique modern minimalist texture can be perfectly interpreted with material features of white lacquer and gray glass.

In addition, the space constraint sense can be through the use of minimalist lines. The designer removed the obstruction of doors and walls in the space with small area through the use of mirror and gray glass, thereby amplifying visual effects. At the same time, the window view design of the chamfered surface is used to add more elegant and meticulous features to the study space. The beam is arranged on the original living room, which is cleverly solved by using streamline curvature of the line board, thereby returning the space to the quiet and simple origin.

HOME
DÉCOR

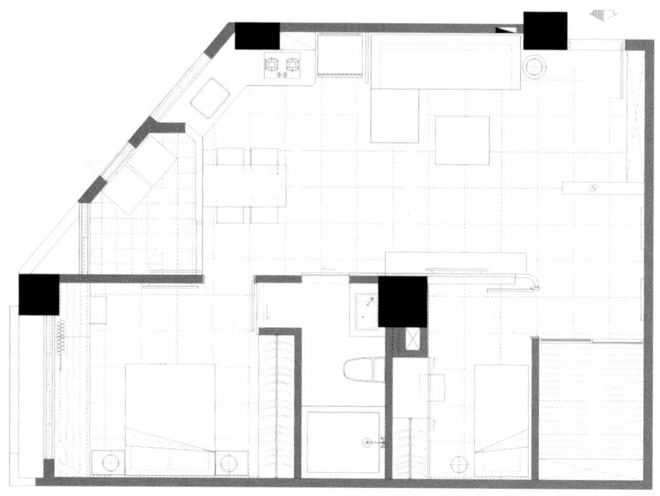

在本案中，设计师让设计更贴近生活习惯，没有夸张的繁复装饰，将空间化繁为简，选用白色为主基色，藉由个性家具的利落线条，刻画出独具个人魅力的风格表情，透过白色烤漆和灰玻璃的材质特性，完美诠释独一无二的现代简约质感。

另外，通过运用极简的线条，解放空间局促感。设计师透过镜面和灰玻璃的运用，让小面积空间少了门片墙面阻隔，放大了视觉效果。同时，利用斜切面的窗景设计让书房空间增添了几分高雅细致。而原本客厅上方存在的大梁问题，也运用线板的流线弧度巧妙化解，让空间回归最沉静简洁的原点。

设计公司：深圳三米家居设计有限公司 | 设计师：三米 | 项目地点：中国广东 | 项目面积：65 平方米

SMALL DWELLING

小蜗居

This is a house with small area, the so-called "sparrow may be small but complete" refers to it. In the space with only 65 square meters, the layout of two bedrooms and two living rooms are not significantly crowded, but shows a bit simple and compact refreshing feeling. Blank wall space is combined with white cabinets to make the space full of stylish simplicity taste; the doors of the kitchen and bathroom adopt invisible doors films, thereby ensuring completion and unity of the public space.

In the matching aspect of furniture soft decoration, in order to meet the space neat lines, simple and elegant furniture and soft assembly decoration with much texture allow space to reveal fashion flavor, elegant colors are matched to support homeowner's personality and temperament.

The living room and dining room are based on expressing simple and stylish features, bedroom design has the theme of leisure, freedom and comfort, fashionable and simple furniture are combined with comfortable bay window design, thereby making the bedroom become the best place for sightseeing.

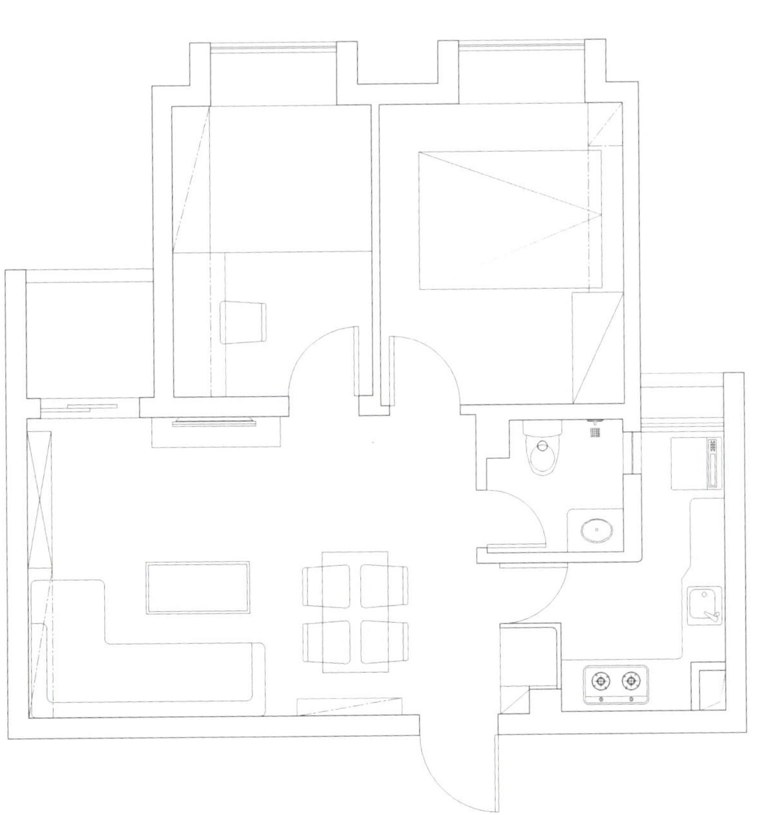

这是一个小面积住宅，所谓的“麻雀虽小五脏俱全”就是这样的吧。在这个只有65平方米的空间里，两房两厅的布局却一点也不显拥挤，反显出几分简约、紧凑的清爽感。留白的墙面结合纯白的橱柜让空间充满时尚简约的味道，而且厨房和卫生间的门也都采用隐形门片来处理，保证了公共空间的完整统一性。

在家具软装的选配方面，为了配合空间干净利落的线条，简约素雅的家具和极富质感的软装配饰让空间流露时尚韵味，素雅的色彩搭配烘托出房主的个性和气质。

客厅和餐厅以表达简约时尚为主，卧室的设计则以休闲、自在、舒适为主题，时尚简约的家具结合舒适的飘窗设计，让卧室成为赏景怡情的最佳场所。

226–320

THE
MININATURE
APPARTMENT
迷你型
公寓

设计公司：鸿扬集团陈志斌设计事务所 | 设计师：陈志斌 | 项目地点：中国湖南 | 项目面积：40 平方米 | 主要材料：银镜、法国木纹石材、墙纸、银箔、水晶帘、地毯

THE SAMPLE FLAT OF WANBOHEI APARTMENT'S TYPE C

万博汇公寓 C 户型示范单位

The house style structure of the case is divided into three sections, the living room and bathroom are located at the entrance, the bedroom is in the middle, and kitchen and space for washing and dying clothes are arranged southwards. Although only 40 square meters are available, but the small space seems to be big, a parlor, a wine cabinet, combined bar and dining table, a shoe rack, a cloakroom, an independent water closet and independent bath are compactly and practically designed, meanwhile, bookcases and desks belong to complete and practical types.

The case is designed with male theme as concept, and exclusive fashion living space is created aiming at fashion male youth in the city. Male themes tend to business functions and concise lines which are tall and straight. The stone texture penetrates throughout, gray mirror is both the wall and the backdrop, and it is also the free creator of virtual and real space. The desk with business function and computer desk are connected for easy operation, it is even always connected with the dining table (bar), which is smooth and natural, comfortable and straight fabric bed is harmonious with soft package background, thereby highlighting the masculinity and charm.

本案的户型结构分为三段，进门处为客厅和卫浴，中段是卧室，靠南为厨房和洗晒衣物空间。虽然只有40平方米，却以小见大，紧凑实用地规划了会客、酒柜、吧台与餐台综合体、鞋柜、衣帽间、独立坐厕、独立沐浴，而且书柜和书桌也是齐全而实用的类型。

本案是以男性主题为理念设计的，针对的是都市中的时尚男青年，为之打造专属的时尚生活空间。男性主题偏向商务功能，线条简练、挺拔方正。石材的质感贯穿前后，灰镜在墙面既是背景墙，又是虚实空间的自由创造者。商务功能的书桌、电脑桌连接起来，方便操作，甚至一直与餐台（吧台）连接一体，流畅自然，舒适而方正的布艺床与软包背景和谐统一，凸显男性气质与魅力。

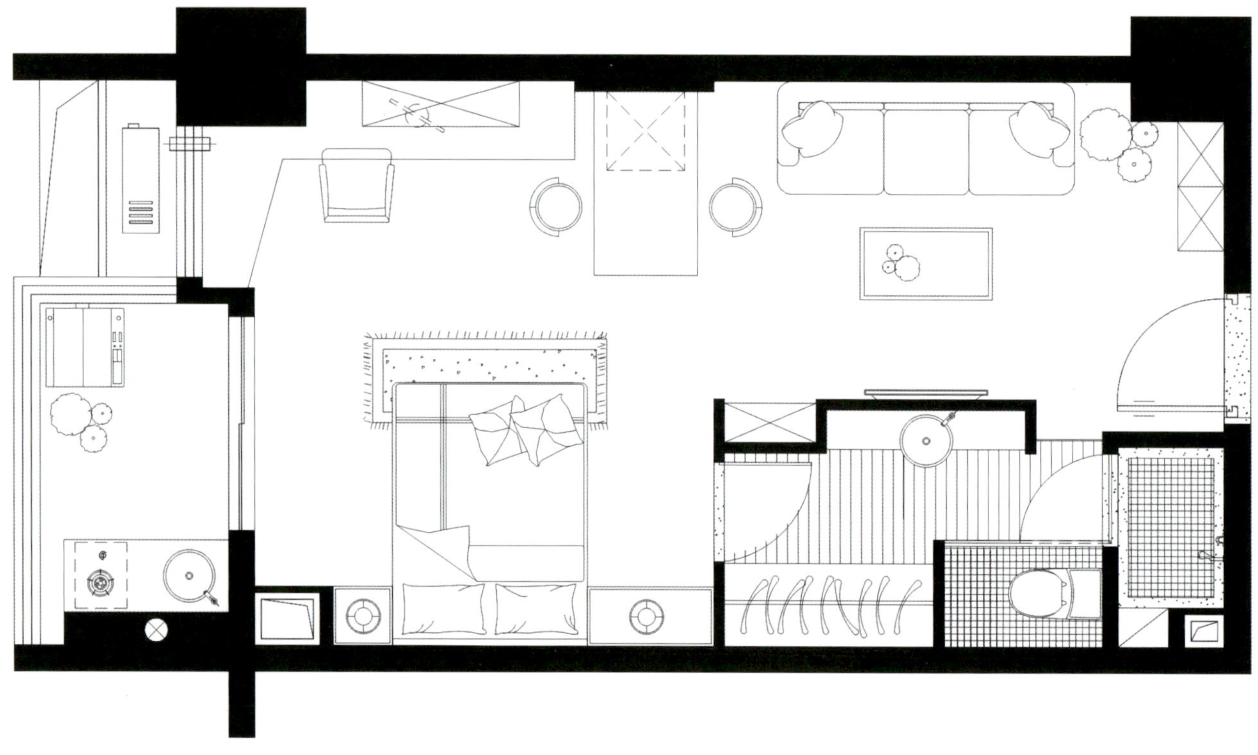

设计公司：鸿扬集团 陈志斌设计事务所 | 设计师：陈志斌 | 项目地点：中国湖南 | 项目面积：40 平方米 | 主要材料：银镜、法国木纹石材、墙纸、银箔、水晶帘

THE SAMPLE FLAT OF WANBOHEI APARTMENT'S TYPE C

万博汇公寓 C 户型示范单位 – 女性主题

The case is an apartment under operation of Hongda Group in urban central area, the female theme is used as the practice of concept, the designer starts from the taste concept of single female to create soft, mellow and romantic atmosphere.

The size structure is divided into three sections as men theme space, including the living room and the bathroom door at the entrance, bedroom in the middle section, and rear kitchen and laundry space. The space with 40 square meters is divided into complete functional areas, meanwhile, fashion and romantic atmosphere is emphasized.

Female theme focuses on home function, round living room and curved sofas are preferred by house woman, soft and romantic crystal bead curtain and chandeliers dot out the fine and warm theme. Combined bar and dining table subtly become transition between the living and dining rooms. Silver lace furniture and pattern wallpaper background echo the romantic theme to dot colorful soft furnishings, thereby an emotional and intellectual space is presented in front of people. Love and none love are shown here for waiting its loving girl's arrival.

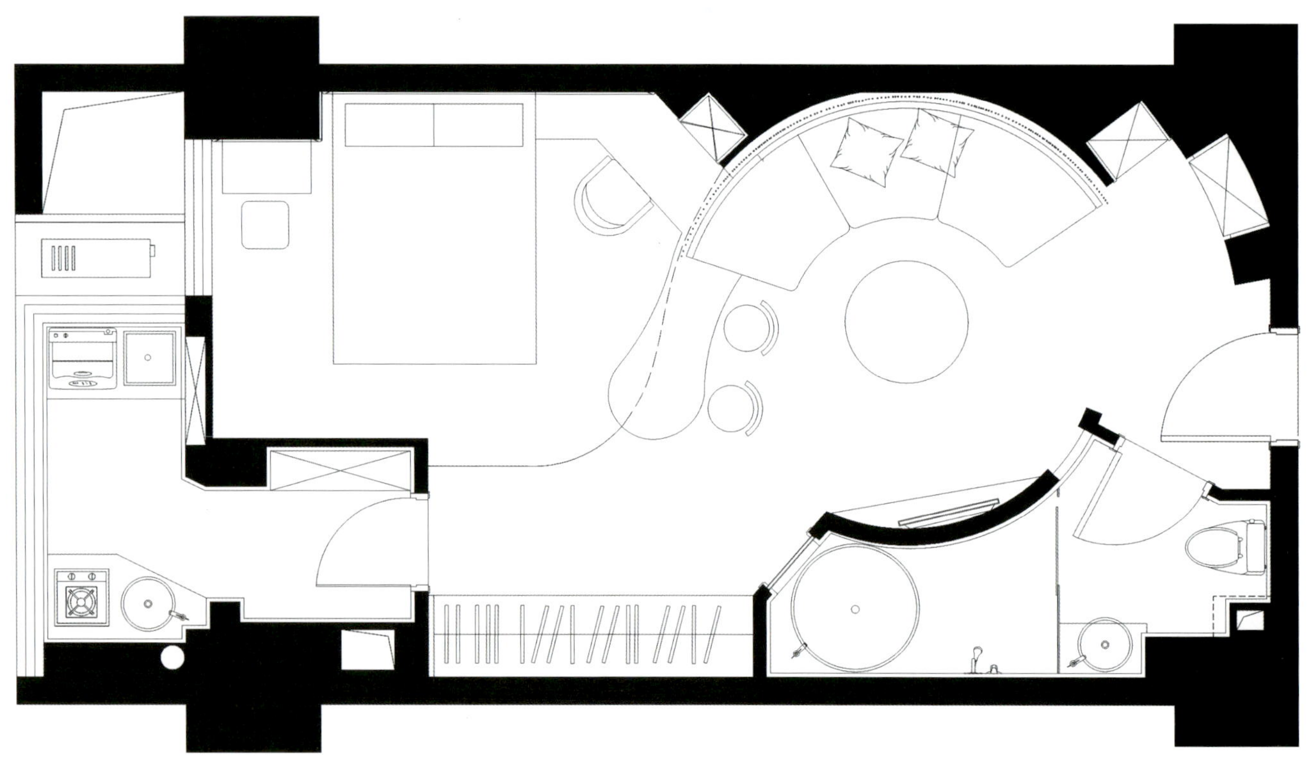

Egypt

本案是建鸿达集团都市央区运营下的公寓，以女性主题为概念的实践，从单身女性的审美观念入手，营造柔和、圆润、浪漫的氛围。

户型的结构与男性主题空间一样分为三段，进门处的客厅和卫浴，中段的卧室，后部为厨房和洗晒空间。40平方米的空间规划出完整的生活功能区，同时强调出时尚、浪漫的氛围。

女性主题偏向居家功能，圆形客厅、弧形沙发是宅女的首选，柔和而浪漫的水晶珠帘和吊灯点缀出精致、温馨的主题。吧台与餐台结合起来，巧妙地成为客厅与餐厅之间的过渡。银箔花边家具与花纹墙纸背景呼应浪漫的主题，点缀丰富多彩的软装饰品，将一个感性而知性的空间呈现在人们面前。爱与不爱，它都在这里，静静等待，钟爱它的女孩的到来。

设计公司：绝享设计工程有限公司 | 设计师：唐俪文 | 项目地点：中国台湾 | 项目面积：56 平方米 | 主要材料：秋香木、喷漆、亮面塑铝板、人造石、云竹地板

TAIPEI-JINGZHAN

台北京站

Only the bedroom and study have daylighting among all room, thereby easy causing dark and close feeling, additionally, the pattern must be broken in order to plan two bedrooms and two living rooms in the apace less than 60 square meters, thereby the designer firstly remove the partition of the master room, the light source of the bedroom can be introduced into the living room and the dining room through the design of arenaceous glass cabinet, meanwhile, close sense of the real wall can be avoided at the same time.

In addition, in order to make the indoor space more bright, whole room takes white as base from porch shoe ark to the TV wall in the sitting room, continuous white plane is matched with the upper and lower indirect light for pavement, thereby amplifying the visual effect, it is matched with black grid adornment ark on the TV wall, which seems to be more plump. In addition, because the structure column and s conduit room are available behind hoe ark, the designer uses cant ark for decoration, and deep and shallow cabinets are planned inside, thereby utilizing the scattered space rationally, and increasing the storage space.

The focus with the most design tension is the red sofa background wall and the white cloth sofa, which dash forward youth and vigor, and brighten space theme.

全室仅有卧室及书房面采光，易造成阴暗、封闭感，加上不足 60 平方米的空间要规划出二房两厅，在格局必须有所突破，于是，设计师先将主卧室隔间拆除，改以喷砂玻璃置物柜的设计，使卧室的光源透入客、餐厅，同时避免实墙带来的封闭感。

另外，为了让室内更显明亮，全室以白色为基底，从玄关鞋柜延续至客厅电视墙，运用连贯的白色平面搭配上下间接光的铺设，放大了视觉效果，再配合电视墙上黑色格子柜装饰，更显丰满。此外，因鞋柜后方有结构柱及管道间，设计师运用斜面柜做修饰，并在内部规划深浅柜体，既合理利用了畸零空间，又增加了收纳空间。

而最具设计张力的焦点莫过于红色的沙发背景墙和白色的花布沙发，突显青春活力之余，也点亮了空间主题。

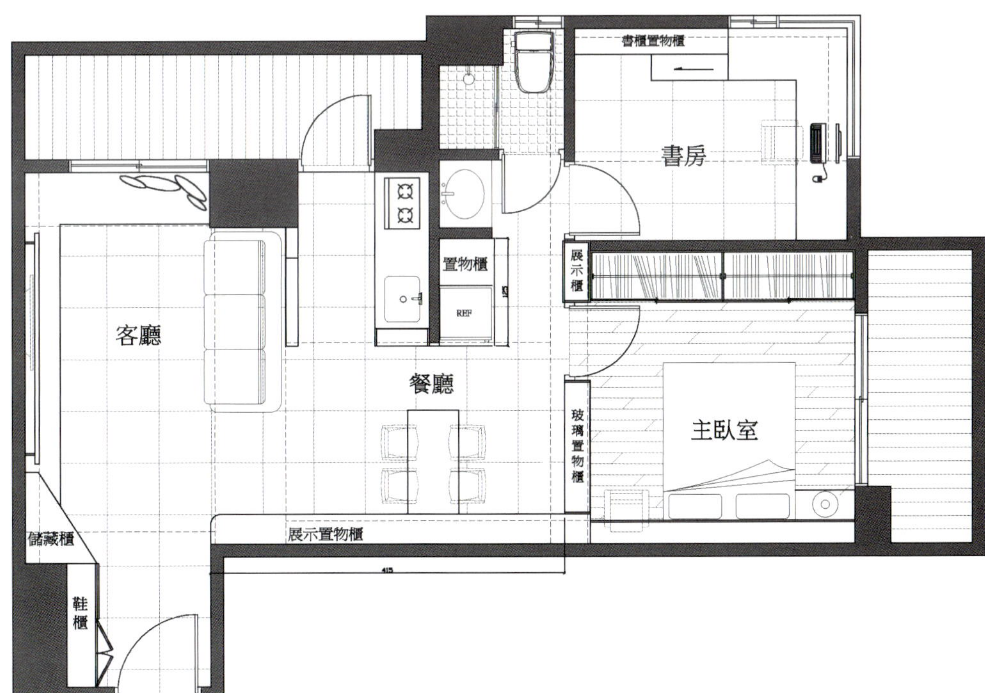
書櫃置物櫃
書房
置物櫃
REF
展示櫃
客廳
餐廳
主臥室
玻璃置物櫃
展示置物櫃
儲藏櫃
鞋櫃

设计公司：界阳 & 大司设计 | 设计师：马健凯 | 项目地点：中国台湾 | 项目面积：88 平方米 | 主要材料：毛丝面不锈钢、铁件烤漆、亮面喷漆、舒然板、钢刷梧桐木

YONG HE YANG RESIDENCE

永和杨宅

This project is located in Xin Dian Xi Tian Guang Scenery with a good position, while there is the disadvantage of lower height of stories and too big beams, so the designer uses the black mirror and the white mirror to cover beams, increases the vision of space height by means of reflected mirrors, and builds a real sense of fashion and clean-cut compared with the baking-paint-glass TV wall below. Meanwhile, the designer skillfully connects the curved line of fixing beams with the ceiling and the TV wall to build a sense of extension united into one, and matches with the carefully calculated triangle TV cabinet. Under the adornment of the opposite sofa wall grey mirror and the light belt, spacious and modern life starts here.

Different from the modern fashion design direction, the major bedroom is changed by the white cultural stone and sycamore wood with natural touch to reveal a humanistic image. The designer locks the lightweight TV set and the audio-visual equipment on the steel column to keep the lightness and completion of the space. At last, a comfortable lounging chair is placed by windows, and the house owner, being a doctor, relaxes his tense nerve, and quietly goes into beautiful time brought by natural scenes.

本案坐拥新店溪天光美景的地利优势，却有层高过低和梁柱过大的问题，设计师以黑镜和灰镜包覆梁体，以镜面反射的层次挑高视感，与下方接续的烤漆玻璃电视墙对比出时尚利落的质感。同时，设计师还巧妙地用修梁的弧线串连起天花板与电视墙面，塑造一体成形的延伸感，搭配精算过动线比例的三角电视柜，在对面沙发墙灰镜及光带的衬饰下，开启敞阔现代生活。

有别于现代时尚的设计方向，主卧室改以白色文化石与手感自然的钢刷梧桐木呈现人文意象。设计师将轻薄的电视与视听器材锁于钢柱上，以保证空间的轻盈完整，并在天花处以间照及造型光带收梁，最后于窗边置一张舒适的躺椅，让身为医生的屋主放松紧绷的神经，静静融入自然光景带来的美好光景中。

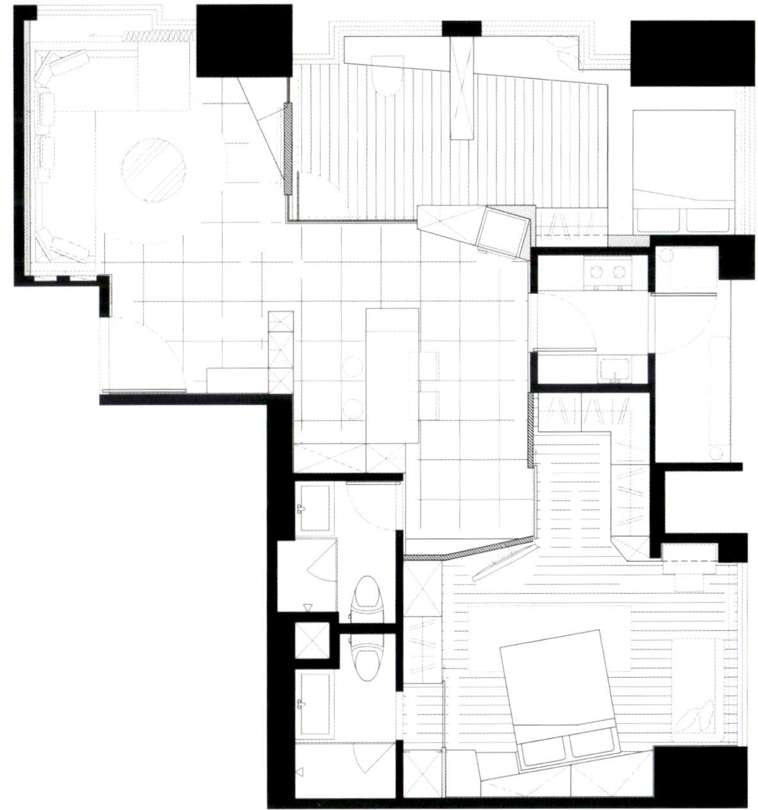

设计公司：德力设计 | 设计师：陈坤海、吴佩瑋 | 项目地点：中国台湾 | 项目面积：29 平方米 | 主要材料：秋香木、黑色板岩砖、马赛克砖、印度黑大理石、灰镜、烤漆玻璃、抛光石英砖、铁件 | 摄 影 师：sam

LOVE AND LOVE TAIWAN HOUSEHOLD PROJECT

恋恋台大住家案

The house is located at the right upper part of MRT station, where it is very convenient for not only transport but also life functions. Because waterproof at the high building and air conditioning problem, the house owner decides to overhaul this project and specially emphasizes to meet the requirement of two people's life space and storage function. Based on above problems and requirements, the designer transforms the originally closed aluminum windows into the adjustable shutters to diffuse heat and help air flowing and prevent rain. In addition, the designer restores the ground row, divides the space into the laundry and clothes drying room, and applies afterheat to dry clothes quickly.

The practical problems resolved, the space becomes more spacious and bright, so people can not feel stuffy even if the space is little. The open kitchen saves space, and change of the bathroom's wall and door makes the space more bright and magnificent. As for materials selection, the plain and elegant materials full of texture are applied to build the space temperament, to make the small space reveal high quality and imposing manner.

戀戀台大施工後平面圖

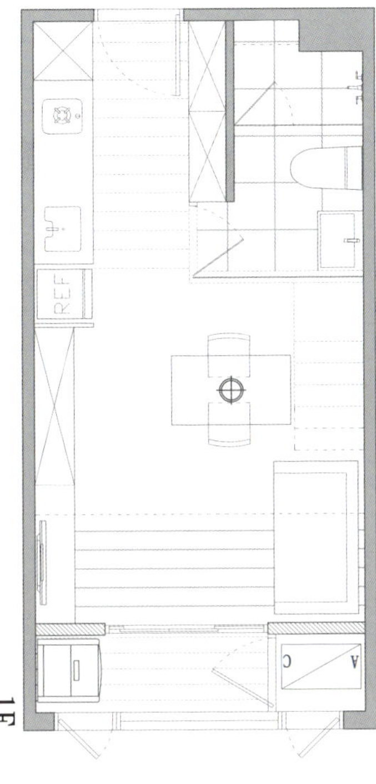

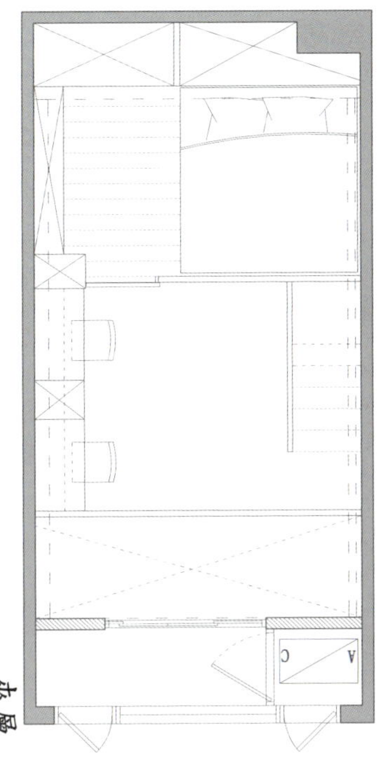

此宅位于捷运站的正上方，无论是交通还是生活机能都极为便利。因为高楼防水和空调不制冷的问题，屋主决定翻修此宅，并特别强调要满足二人生活的空间机能和收纳需求。设计师针对以上问题与要求，将原本的密封铝窗改装为可调整的百叶铝窗，让热气得以扩散，同时有助空气对流和防雨。另外，设计师恢复了地排，并将其空间辟为洗衣房与晒衣间，运用余热快速烘干衣物。

解决了实际问题，空间也变得通透开朗许多，这样一来，即便空间小也不会觉得窒闷。开放式的厨房节省了空间，而卫浴间墙体及门片的改动也让空间更显明亮大气。在选材方面，也通过运用素雅而富有质感的材质来打造空间气质，让小空间流露出高品质、大气魄。

设计公司：贵州峰上室内外设计工程有限公司 设计师：曾涛 项目地点：中国贵州 项目面积：66.84 平方米

主要材料：马可波罗砖、罗马利奥砖、墙纸、西奈珍珠石材、石英石、科勒洁具

GUIZHOU BAILING FASHION WORLD FINE DECORATION APARTMENT MODEL ROOM (VI)

贵州百灵时尚天地精装公寓样板房（六）

This is a design aiming at the young people in a moment with pursuit of individuality and fashion, small size plus perfect design make people easier to accept, but it also can fully meet their psychological needs and economic load capacity.

The case takes 'thorny rose" as design theme, and designer shows the design soul by giving space some temperament. Black is freely swayed here for laying grim personality tone for the space, coupled with the soft furnishings with rich texture and metallic texture item matching, modern fashion is thoroughly displayed. The alligator skin is also used for large area, shiny stainless steel and mirror shadows are whirling and beautiful, crystal chandeliers are so delicate with so costly colored glaze, it is a cool and pretty apartment, the distinguished New Zealand dairy cow leather is also willing to worship a single step.

In the times with hurried pace, the black shows full coolness, and blurred fashion so that more and more people with the need to listen to the inner calling and heart desire at the bottom to find a resonance, this is the "thorny rose" with individuality and creativity.

YIHONG

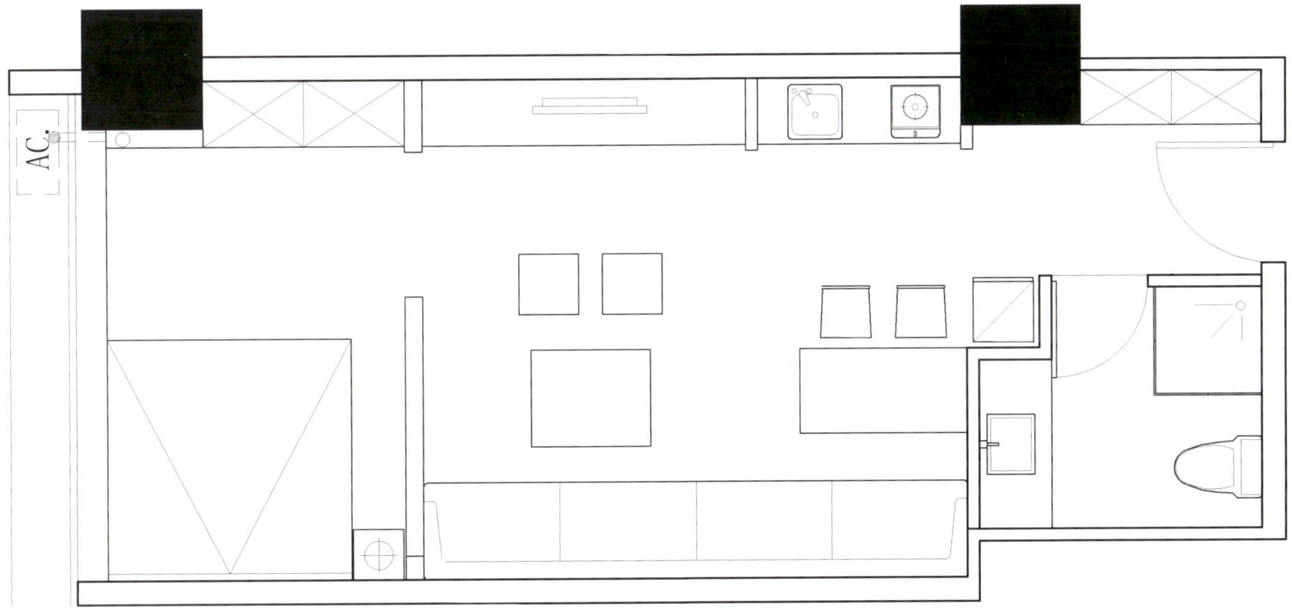
AC.

这是一个针对当下追求个性与时尚的年轻人的设计，小户型加上完美的设计让人比较容易接受，同时也能充分满足其心理需求及经济负荷能力。

本案以“刺玫瑰”为设计主题，设计师通过赋予空间某些气质来展现设计的灵魂。黑色在这里自如挥洒，为空间奠定冷酷个性的基调，加上富有质感的软装饰品与金属质感的物品搭配，将现代时尚表现得淋漓尽致。鳄鱼皮也得到了大面积的运用，不锈钢闪闪发光，镜面倩影婆娑，水晶吊灯如此精美，琉璃饰品那么娇贵，如此酷而俏的公寓，尊贵的新西兰奶牛皮也甘拜足下。在这样步伐匆忙的时代，黑色呈现出的冷感十足且迷离时尚让越来越需要倾听内心呼唤与心底渴求的人们找到共鸣，这就是个性而有创意的“刺玫瑰”。

设计公司：贵州峰上室内外设计工程有限公司 设计师：曾涛 项目地点：中国贵州 项目面积：57.83 平方米

主要材料：马可波罗砖、罗马利奥砖、墙纸、西奈珍珠石材、石英石、科勒洁具

GUIZHOU BAILING FASHION WORLD FINE DECORATION APARTMENT MODEL ROOM (III)

贵州百灵时尚天地精装公寓样板房（三）

This set of single apartment creates graceful rolling romantic atmosphere with soft vine smooth curvature of ups and downs. The same narrow space makes the designer people ignore the original structure through clever planning of the designer, the owner can be only immersed in a wonderful smooth space. The unique big round bed of the bedroom appears immediately when you pull the screen-like curtain. The living space integrates many elements and integrates bath, parlor, dining bar and saloon into a whole, and the clever connection of complex functions make the small space to accomplish a great deal.

The small space also needs luxury temperament, designer creates a unique luxury in the apartment with small area, and people ignore the space area problem, which is indeed painstaking. The designer takes "single multi- tone" as design theme, and creates the colorful living space for a single white-collar worker in the city, and monotonous life is full of fun mean. So, the materials with rich texture and the circular space layout are justified.

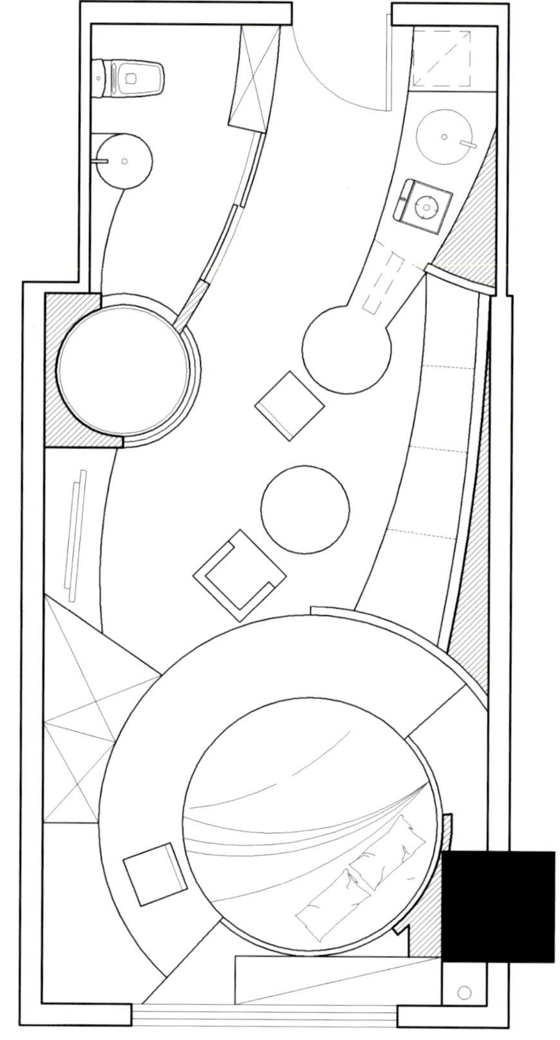

这套单身公寓以柔蔓起伏的流畅弧度，营造婉约连绵的浪漫氛围。同样是狭长的空间，通过设计师的巧妙规划，让人忽略掉原本的结构，只沉浸在美妙流畅的空间中。拉开以为是窗幔的帘子，卧室独特的大圆床立马呈现。起居空间多元一体，集沐浴、会客、餐吧、酒吧多用于一身，复合功能的巧妙衔接，使不大的空间大有作为。

小空间同样想要奢华气质，设计师在这个小面积公寓中营造一份独有的奢华，并让人忽略掉空间面积问题，确实是煞费苦心的。设计师以“单身多调”为设计主题，为都市里的单身白领们打造一个有滋有味的生活空间，让单调乏味的生活充满情趣意味。如此，富有质感的材料与圆形的空间布置都各有道理。

设计公司：焦艳丰设计公司 | 设计师：焦艳丰 | 项目地点：中国重庆 | 项目面积：37 平方米 | 主要材料：科勒洁具、法国木纹大理石

QIAO YUAN SHI JIA

桥苑世家

What this project shows is the unique small space being the design. The space of 37 square meters uses the design concept of free-fall curve from Mondrian Piet. The designer begins from furniture, and defines the space as modern style, where the whole space shows a curved shape from the furniture to the wall, offering people a gentle and graceful feeling.

What this space applies most is the mirror material, to meet life requirement of the house owner, and to have a great effect on enlarging the space. The completely open design is adopted on the reconstruction of the space. Meanwhile, to meet the house owner's high-quality requirement of sleeping, the designer specially chooses the high-quality rounded bed to help sleep, not abrupt in a small space, but more harmonious with the space style.

In addition, what is worth mentioning is the design of the dining room, where the table layout chooses the tree trunk, leaves and other natural elements, adding a highlight to this originally modern and fashionable space. It may be said that small as the sparrow is, it possesses all its internal organs.

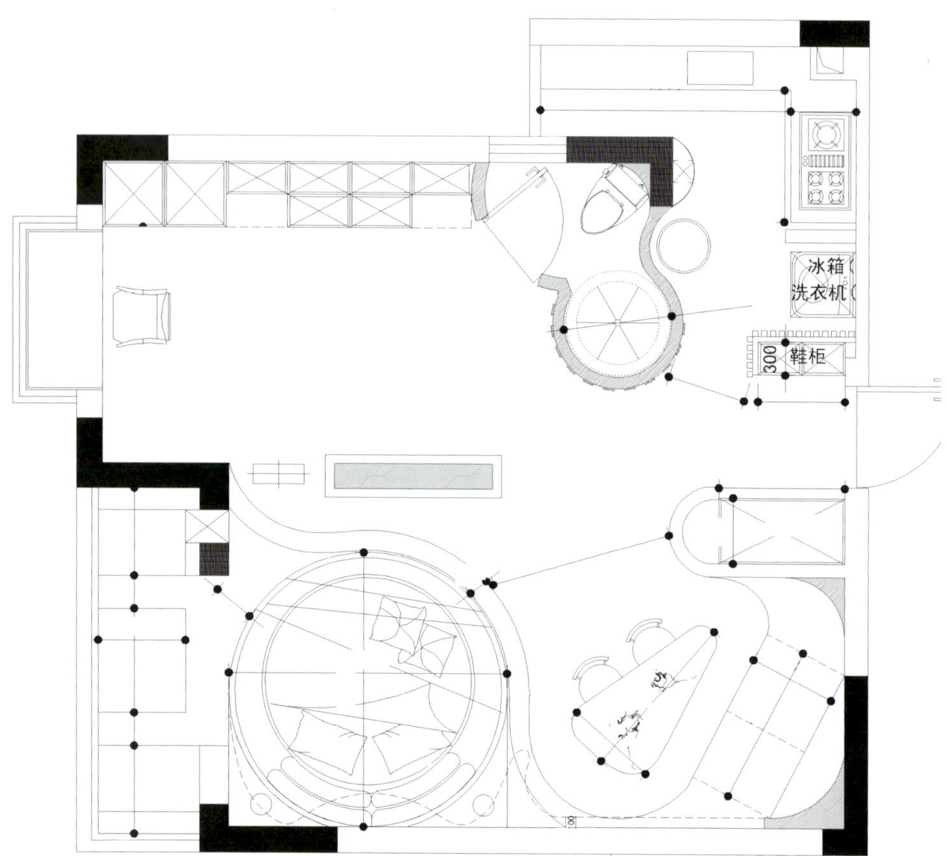

本案展现的是独特的小空间大作为设计。37 平方米的空间采用了平面大师蒙特列安自由曲线落体的设计理念，设计师从家具开始着手，将空间定位为现代风格，整个空间从顶面家具到墙面均呈曲线形态，给人柔和飘逸的感觉。

本案空间运用最多的材料便是镜片，以此满足业主的生活需求，同时也有放大空间的效果。在空间的改造上采用完全开敞的设计，同时，为了满足业主对睡眠的高质量要求，特别选用有助睡眠的高品质圆床，在小空间里也不显突兀，反倒更协调地与空间风格融合在一起。

另外值得一提的是餐厅的设计，餐桌布局选用了树干、树叶等自然元素，在原本的现代时尚的空间中平添添一笔亮色。，可谓麻雀虽小五脏俱全。

设计公司：香港方黄建筑师事务所 设计师：方俊 项目地点：中国重庆 项目面积：46.19 平方米

CHONGQING REN AN A1

重庆仁安 A1

This is a small residential area, whose target customers are defined as single youngsters. The designer takes GUCCI brand clothing into consideration, and reflects such a kind of thinking way by means of colors, materials and the light source. This concept is suitable to every way, so the designer adopts the clothes element to the interior design. This is a kind of consideration from the designer, and also a fashion reflected by this project design.

The room space is small, so there is nothing about the space layout planning, but the design in details must be original. The enclosed bathroom at the entrance has the space avoid taking in everything in a glance, the kitchen and the dining room show themselves along the passageway, opposite the dining room is warm and exquisite small living room, and a hollowed-out iron screen is set between the living room and the sleeping bed, all of which are a kind of artistic expression, a kind of fun expression way. The innermost part of the space is the edge of the sleeping bed by windows with a work desk, where it is not only a leisurely place to view the scenery, but also the desk to study and work.

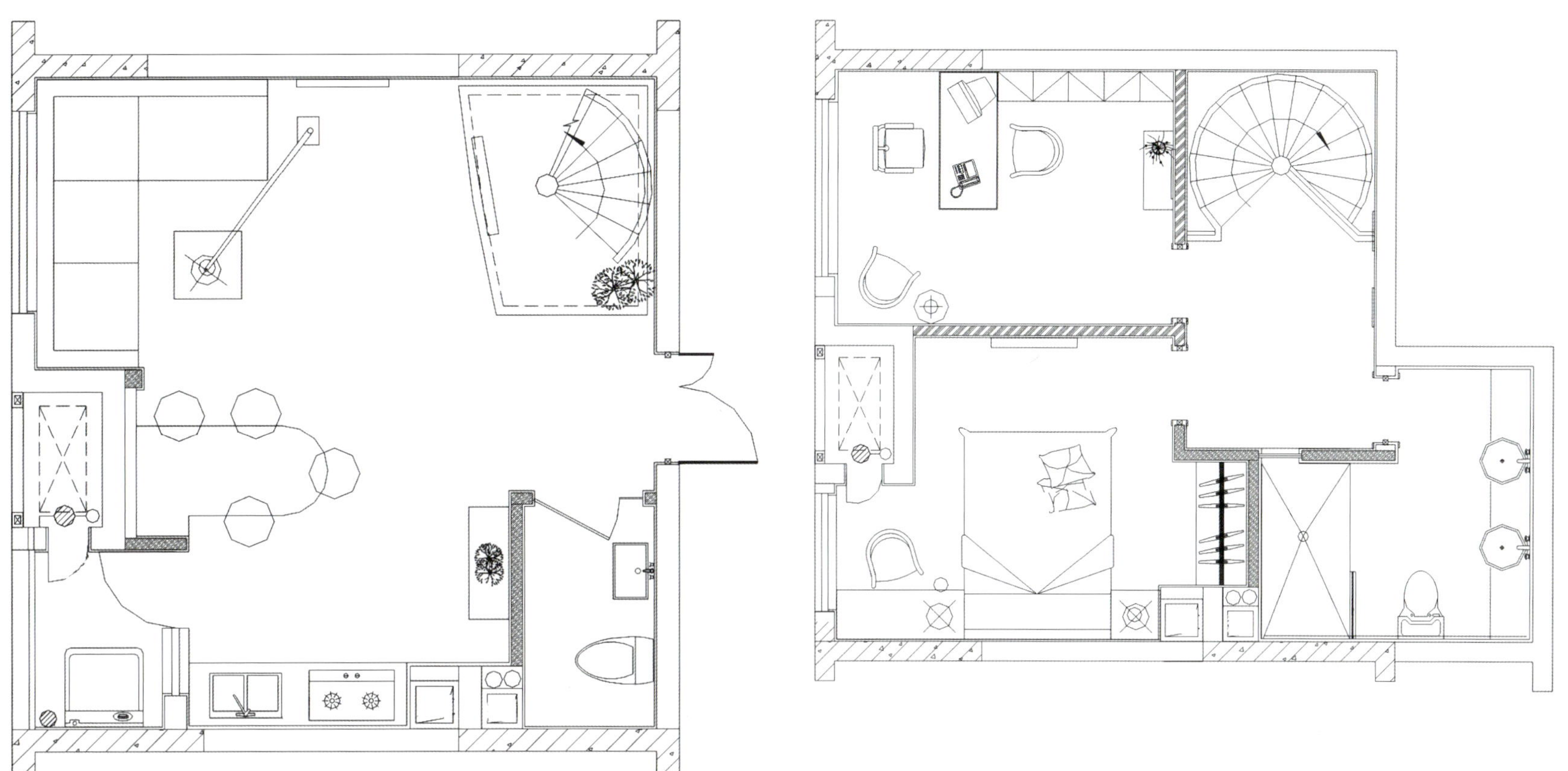

这是一个小面积住宅，目标客户定义为都市单身青年，设计师藉由 GUCCI 品牌服装为思考，通过色彩、材料、光源三方面来体现这样的一种思考方式。设计师共通的，也是共融的，于是，服装的元素同样可以运用在室内的设计中，这是设计师的一种考虑，也是在本案设计中体现出来的一种时尚。

房间面积很小，谈不上空间格局的规划，但是细节上面的设计却必须独到。入口处藉由封闭式的卫浴间让空间避免了一览无余的境况，厨房和餐厅沿着走道排开，餐厅对面是温馨精巧的小客厅，客厅与睡床之间隔着一道镂空的铁艺屏风，这是一种艺术的体现，也是一种情趣的表达方式。空间最里面，也就是睡床的边缘，靠窗八方了一张工作台，既是休闲赏景的去处，也是学习办公的台面。

CHONGQING REN AN A2

重庆仁安 A2

The model room is a small - size apartment which is specially designed for the young, thereby the designer has the theme of “fashion and avant-garde" and obtains inspiration and elements of design creation from popular fashionable dress. The designer selects the fashionable dress of brand GUCCI to extract the elements, which illuminates the understanding and control of designer for design aesthetics and the understanding of leading the trend of the times and popular classic contained in fashion and avant-garde.

Although the case has small area, the idea is reflected through the creative thought of the design, extracted design elements, various building materials, mastering abilities of display articles, and the detailed elements of the corresponding model room can be observed through each set of fashionable dress. Careful and rarefied selection and use of material make the room exude a touch of natural breath. In these delicate little spaces, the professionalism of the designer gives the space extremely reasonable and practical function partition layout, so that living can distribute beautiful and comfortable temperament from inside to outside.

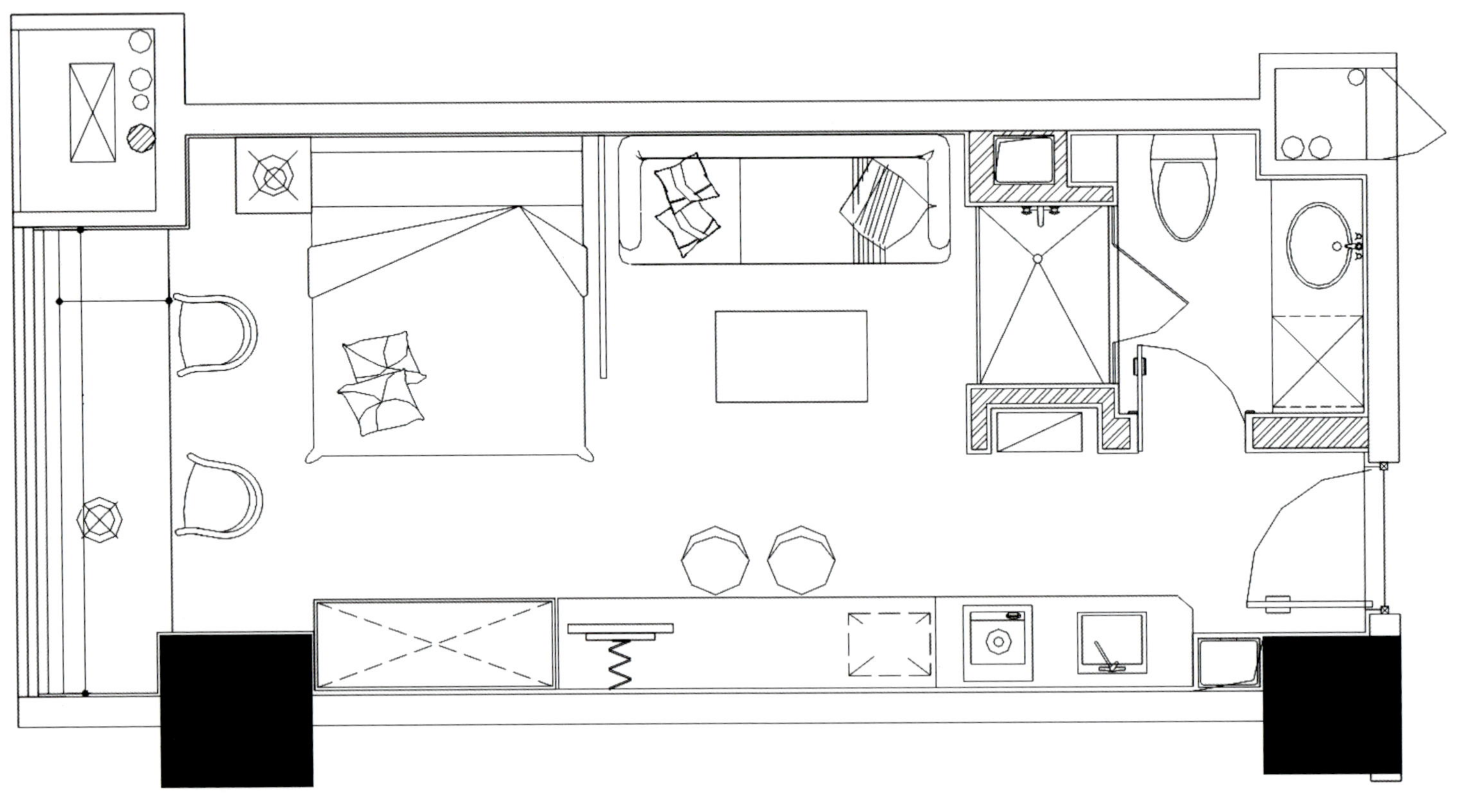

该样板间是针对年轻人而专门定制的小户型寓所，因此设计师以“时尚、前卫”为主题，大胆地从当下流行的时装中获取灵感及设计创作的元素。而选择 GUCCI 这个品牌的时装提取元素，则说明了设计师对于设计美学的理解与把控以及对于时尚、前卫中蕴含的“引领时代潮流与流行经典”的理解。

本案面积虽不大，但理念的东西却通过设计师创造性的思维及对提炼的设计元素与各种建材与陈设物品的驾驭能力，透过每一套时装看到相对应的样板间的细部元素。材料精心考究的选取与运用让房间散发着淡淡的自然气息。在这些精巧的小小空间里，设计师的专业素养赋予了空间极其合理与实用的功能分区布局，让居住由内而外散发美丽、舒适的气质。

设计公司：怀特室内设计 | 设计师：林志隆 | 项目地点：中国台湾 | 项目面积：82.5 平方米 | 主要材料：清水板模漆、烤漆玻璃、樱桃木、美耐板、不锈钢

45 DEGREES

45 度

This is an integrated space of work and life, skillfully combining living space with personal studio, which not only emphasizes the human nature and performance, but also reveals its strong space feature. Therefore, in the studio area, the designer adopts modern fashion as the design fundamental key, and outlines the pure space layers without color. Because of low ceiling space, a great deal of white is the space partition for space background and grey mirror, which not only extends the space and brightens the vision, but also could make scenes.

Modern and clean-cut fashion space, work table and bookcase are paved by high and low irregular lines, matching with windows mainly with geometry lines and white walls, which immediately bring clear and clean work space with professional quality, as well as bight and spacious unique and exclusive space. As for the living space, the kitchen and the dining room are planned into the same space, and spread-eagle bar extension combined with the dining table increases storage space of the cabinet below, making life being more at ease.

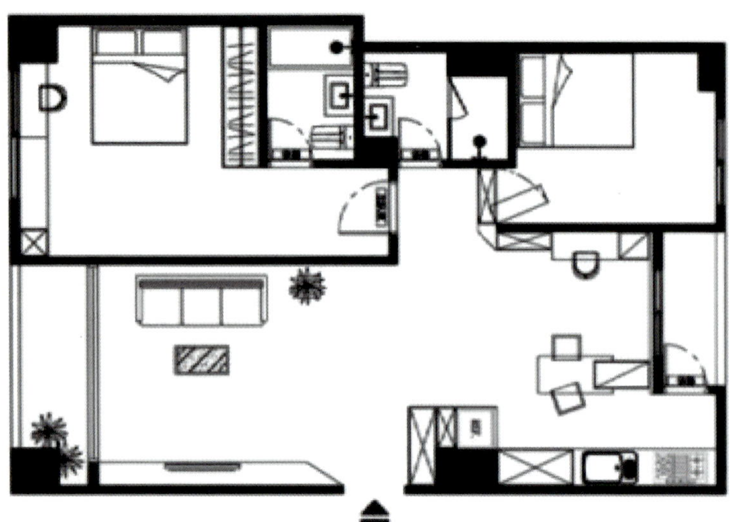

这是一个工作和生活相结合的综合性空间，要将居住和小型的个人工作室巧妙地结合在一起，除要重视人性机能外，还需有浓厚的空间特色。因此，在工作室区域，设计师以现代时尚为设计基调，以无色彩的纯粹来铺陈出空间层次。由于空间天花较低，以大量的白色作为空间背景、灰镜玻璃作为空间区隔，在空间放大及视觉通透之余，也可以制造工作景致。

现代利落的时尚空间，工作桌、书柜均采用高低不规则的线条铺陈，配合上几何线条边框为主的窗户及白色为基调的墙面，立刻带出条理分明、有专业素养的工作区域及明亮开敞的独特专属空间。生活区，将厨房与餐厅规划在同一空间中，利用一字型的吧台延伸与餐桌的结合，增加下柜的收纳空间，让生活更显自在。

White

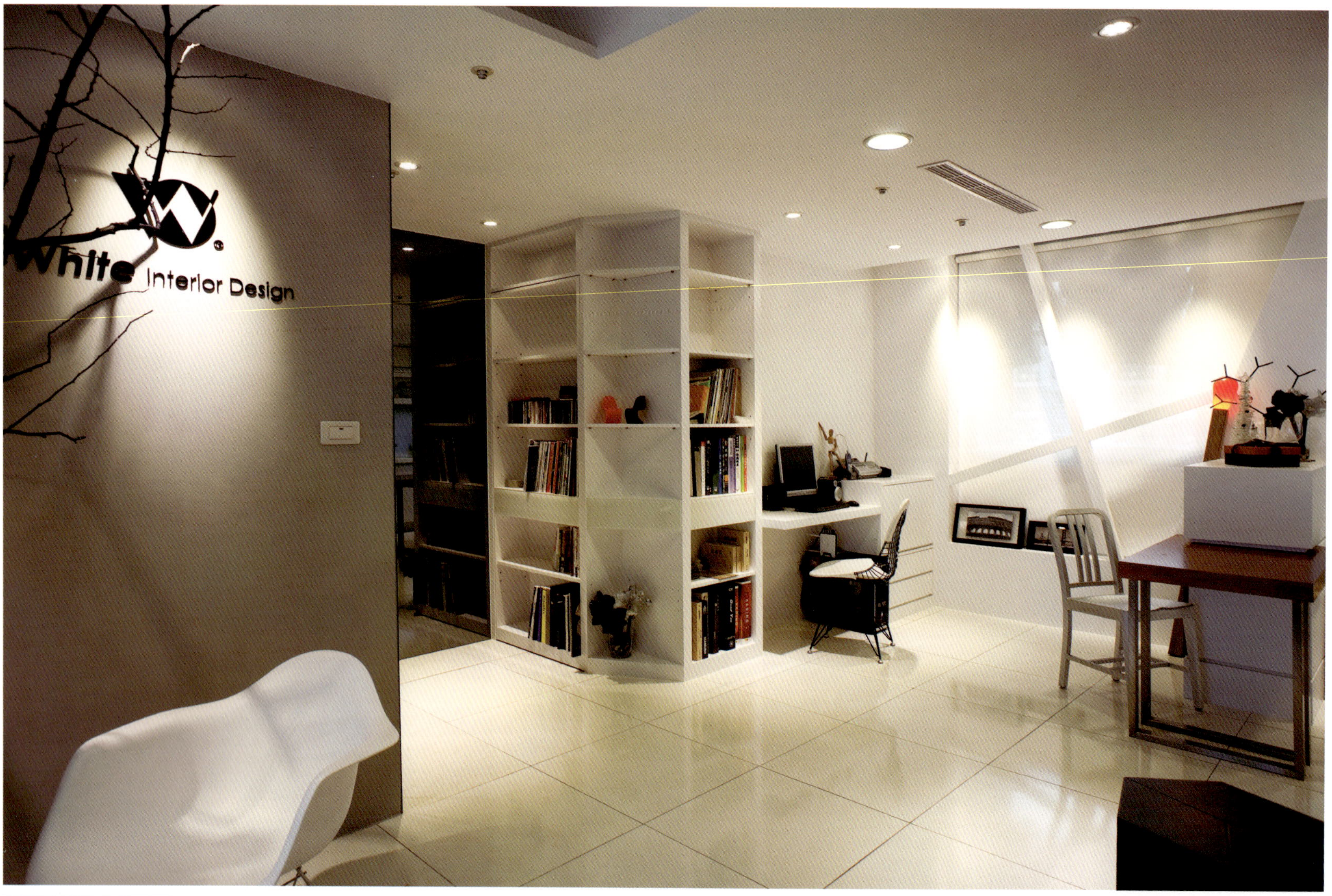

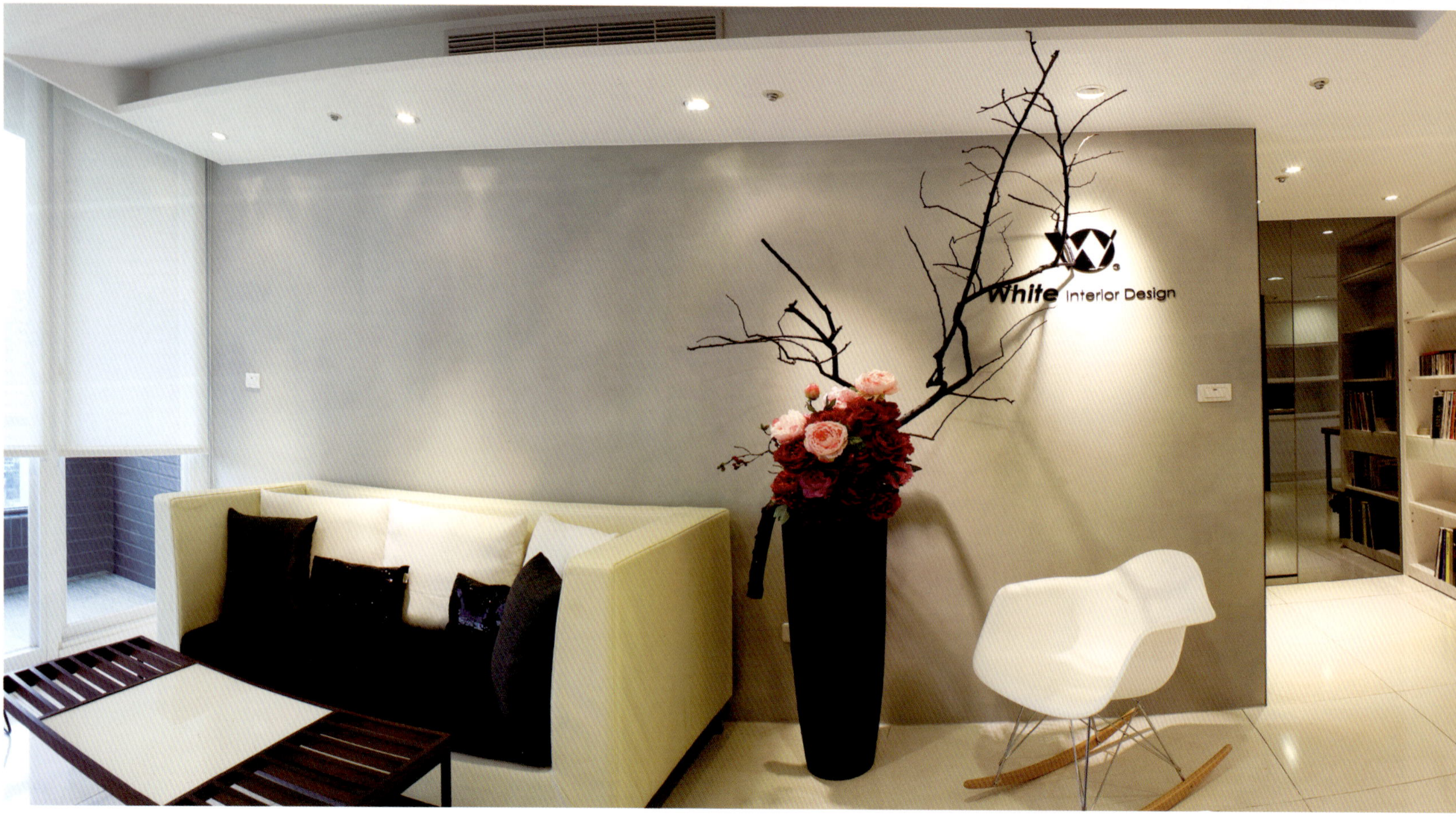

HUI CUI MANSION MODEL UNIT ROOM 01

汇翠山庄样板房 01 房

In this not large space, the designer applies modern and concise style to match with Chinese elements, and presents a plain but elegant "home" to people. Concise space design plus profound Chinese elements make the whole space being serene, fresh, concise but not simple. This is a household space full of implication, bringing people a different life experience.

In the living room, behind the cream-colored sofa is coffee-colored hard packet processing, plus white Chinese partitions and the application of the tea table with Chinese decorative design, making the living room elegant and serene. The open-style kitchen and the living room act in cooperation with the TV setting wall across the white dining table, plus the application of mirror surface, making the space more open and spacious. The design of the bedroom continues the general space style, serene and quiet, and large-glass windows bring the indoor enough lightings, adding a more tinge of warm and fashion.

生活阳台
3.1M²
主卧
12.9M²
书房
7.4M²
卫生间
6.0M²
客厅
20.8M²
餐厅
厨房
13.7M²

Arletta

在这个面积不算大的空间里，设计师运用现代简约风格搭配中式元素，将一个素净优雅的“家”呈现在人面前。简约的空间设计加上饱含深意的中式元素，让整个空间显得沉静、清爽，简约而不简单。这是一个充满意韵的家居空间，带给人不一样的生活体验。

客厅里，米色的沙发后面是咖啡色的硬包处理，加上白色的中式隔断和采用中式纹饰设计的茶几，让客厅显得优雅而沉静。开放式的厨房与客厅隔着白色的餐台和电视墙遥相呼应，加上镜面的运用，让空间更显开阔敞亮。卧室的设计延续着大体空间的风格，沉稳中透着宁谧的气息，大幅玻璃窗让室内光线充足，更添几分温馨、时尚。

设计公司：大匀国际空间设计　室内设计：林宪政　软装设计：毛方方　项目面积：40 平方米　项目地点：中国江苏　主要材料：墙纸、白色喷漆、方块条砖、柚木地板

SMART SPACE FOR SINGLE UPSTART—NANJING CAIYUNJU C DWELLING-SIZE APARTMENT

单身新贵的酷睿空间——南京彩云居 C 户型

This type of house is mainly developed for single superior people, which adopts black and white lines to demonstrate the character, fashion and warmth of the space. People can be smart, cool and chic even though single.

This project is aimed at such design concept to be created. As the space is not large and the structure is the elongated shape which can be seen thoroughly from the top to the bottom, the placement of this space is mainly with clearness and freshness, and the temperament and specialty of the space is served by material texture and colors, to more highlight its space advantages. Here, pure white color tone plus modern concise furnishings first add a sense of fashion and elegance to the space, matching with teakwood flooring, which not only coordinates the integral atmosphere of the space but also adds a sense of warmth to the space, so as to make the single upstarts experience the fashion and feel the warmth and comfort of the small home.

此户型以单身贵族为主要销售对象，设计多采用黑白线条来表现其空间性格：时尚不失温馨的主体特点，即使单身，亦是酷睿潇洒。

本案就是秉着这样一个设计理念来进行创作的，因为空间不大，结构也是一眼到底的狭长型，因此在格局的布置方面以简洁清爽为主，空间的气质和特色则以材料质感和色彩搭配来进行烘托，让空间的优势更加突出。在这里，纯白的色调加上现代简约的家具设施首先为空间添上时尚优雅的一笔，再搭配柚木地板，在协调整体氛围的同时为空间注入温馨感，让都市里的单身贵族在体验时尚的同时，也感受到小家的温馨和舒适。

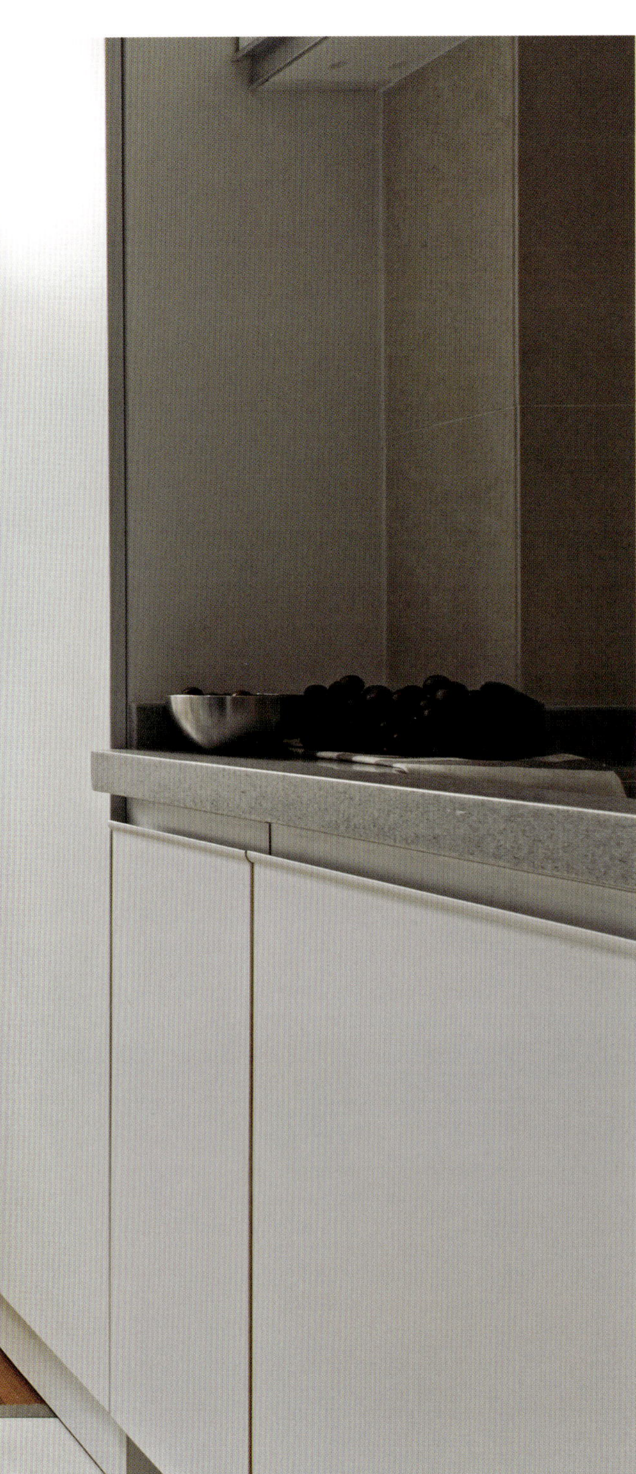

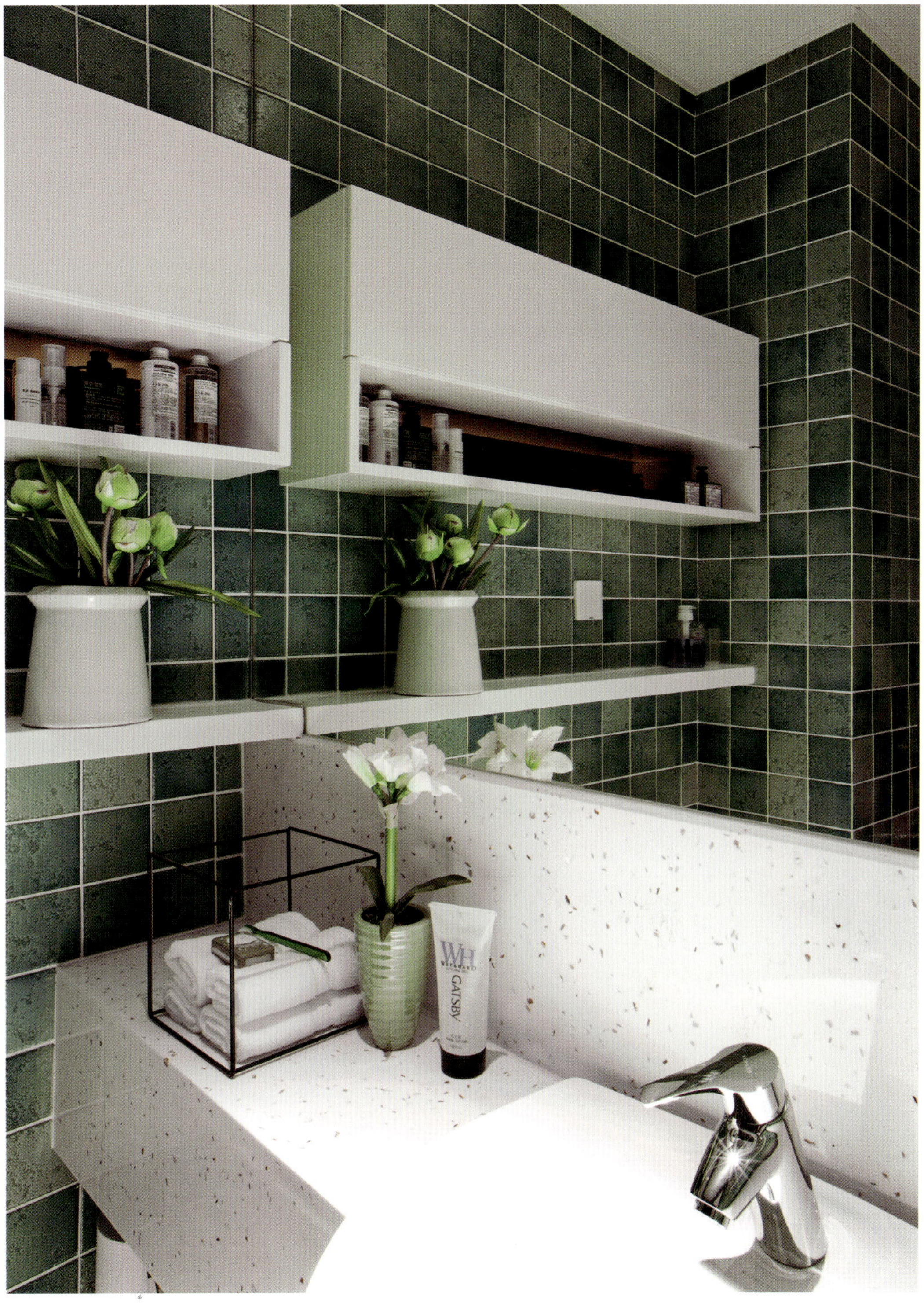

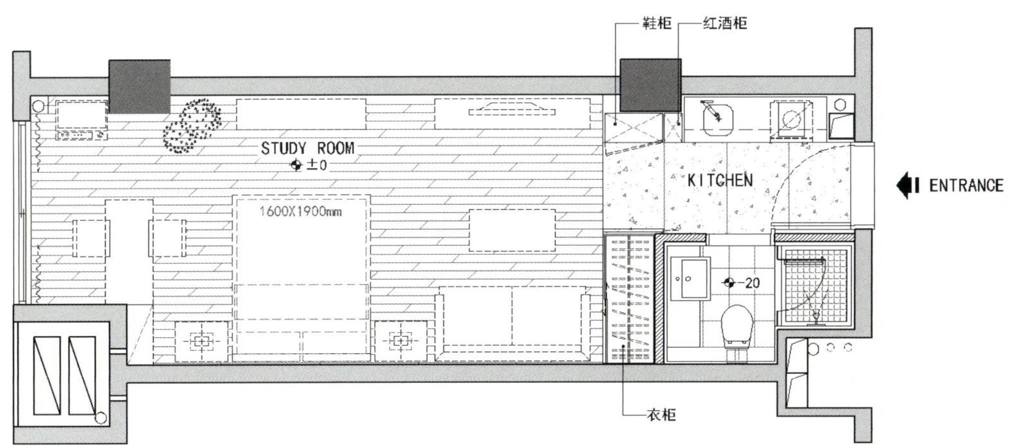

GATSBY

鞋柜
红酒柜
STUDY ROOM
±0
KITCHEN
ENTRANCE
1600X1900mm
-20
衣柜

设计公司：北京百安居设计公司 | 设计师：李源 | 项目地点：中国北京 | 项目面积：48 平方米 | 主要材料：地板、瓷砖、乳胶漆、橱柜

FULL HOUSE

浪漫满屋

In this space less than 50 square meters, the designer wants to create the atmosphere of full house for the homeowner, thereby ignoring the size of the space and focusing on the enjoyment of life.
The small size of house should firstly solve the problem of containing, the designer created a Tatami room at the entrance, raised floor not only solves the admission problem, but also separates from the public areas, the room is created according to the form of Tatami room, hazy and floating feeling can be generated with the form of wire curtain enclosure, meanwhile, colorful features can be added to the house at the same time. The living room, the dining room and the kitchen form trinity, a ceiling is made with vinyl floor tiles among the three parts, which is used for separation, and can activate the space environment. A big bay window is arranged by the window to ensure the lighting and ventilation of living room and the Tatami room.
The bedroom consults the design of the Tatami room; the bed enclosed by the cabinets is also a place for containing articles, which is concise and practical. The purple window curtain responds to the TV background wall of the living room, which is elegant and romantic.

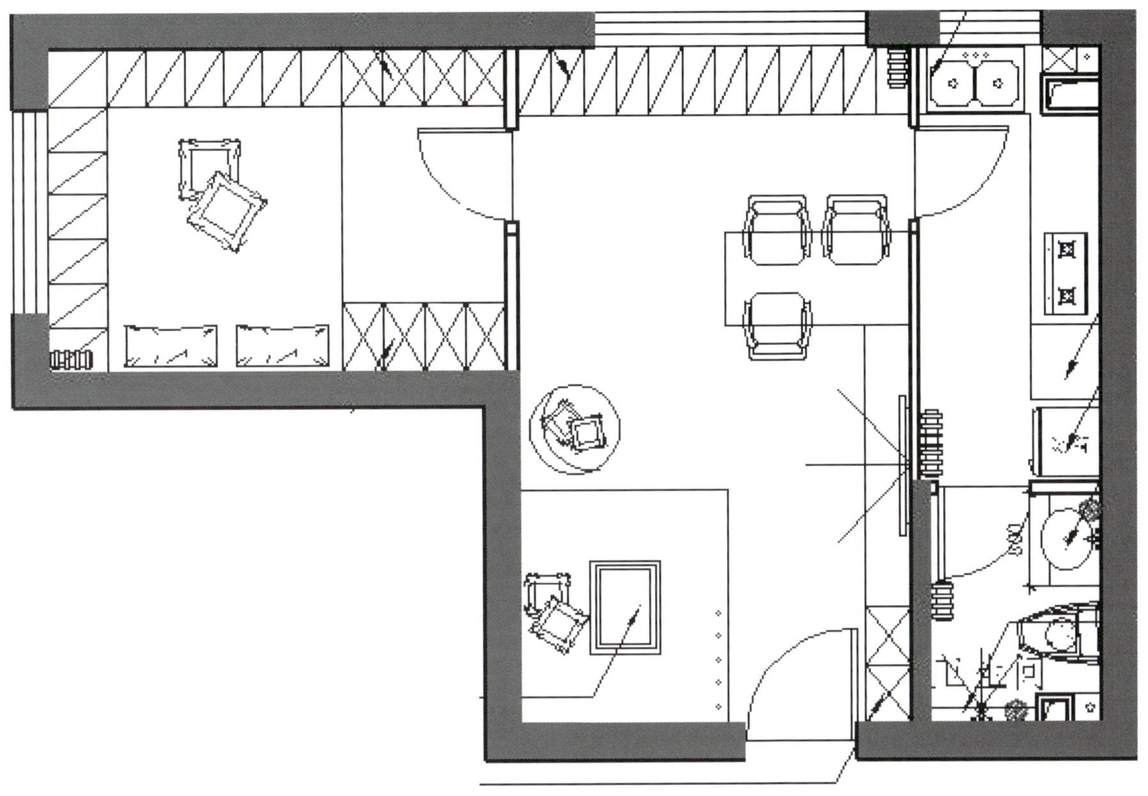

在这个面积不足50平方米的空间里，设计师想要为房主营造出浪漫满屋的氛围，忽略空间的大小，专注于生活的享受。

小面积住宅首先要解决的就是收纳问题，设计师在入口处打造出一个和室，抬高的地板既解决了收纳问题，又与公共空间很好地区隔开来，和室以开放式的形式打造，以线帘围合的方式带给人朦胧飘逸的感觉，同时也多出几分艳丽之色。客厅与餐厅、厨房三位一体，之间以仿木地砖做出一道吊顶，既是区隔，又活跃了空间氛围。临窗的位置有一个大大的飘窗，保证了客厅和和室的光照和通风。

卧室参照和室的设计，以柜体围合起来的“床”同样是物品收纳的好去处，既简洁又实用。紫色的窗帘呼应着客厅的电视背景墙，优雅而浪漫。

设计公司：绝享设计工程有限公司 设计师：操莉玲 项目地点：中国台湾 项目面积：40 平方米 主要材料：雪白银狐大理石、文化石、马赛克、超耐磨木地板

YONGHE NEW ERA

永和新时代

The designer posted a mirror of at the porch of the entrance, a poetry favored by the hostess is written on the mirror, romantic atmosphere is dyed with lighting, and the space of the porch can be amplified while the space can be beautified.

The light of the entire space is only from one side, French door film is matched with frosted glass to introduce light into the sofa wall, a whole culture stone extends into the room, white paint is brushed on the wall with rough texture, and it becomes very tasty. The place of transfer is provided with a shaped wall panel, traditional and simple lines naturally show the classical style, and hallway evil can be avoided. The kitchen is posted with glittering mosaic, the bar area is accompanied by a small crystal lamp, and the dining area atmosphere is suddenly increased.

The master bedroom is set beside the lighting window of the house, and you can enjoy the warm sunshine. Culture stone wall in the room is matched with bedside backplane with stretched skin, classic bedside lamp and short cabinet, thereby dotting out thick Parisian style, romantic style is shown without superfluous decoration.

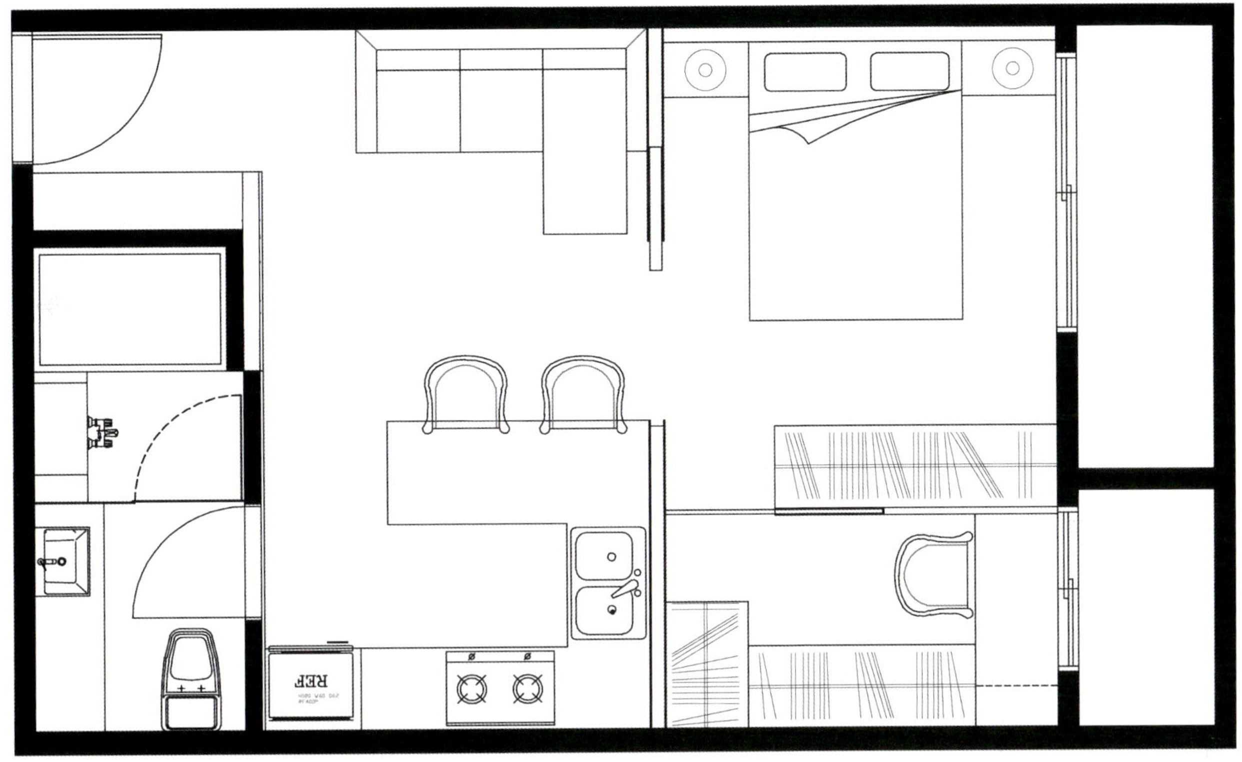

设计师在入口的玄关处贴了大面的镜子，并题上女主人喜欢的新诗，以灯光晕染浪漫氛围，美化空间之余，更有放大玄关空间的作用。

整个空间的光源只有一面，以法式门片搭配喷砂玻璃将光引入沙发墙，一整面的文化石延伸到房间内，带有粗糙质感的墙面，刷上白漆，就变得很有味道。而交接的地方则是造型壁板，传统简约的线条自然地带出古典风味，还避忌了穿堂煞。厨房贴了亮晶晶的马赛克，吧台区配上小盏的水晶灯，用餐区的氛围顿时上升了一个热度。

主卧设置在房子的采光窗户旁，可以享受温暖阳光。房里的文化石墙面搭配绷皮的床头背板、古典的床头灯及矮柜，点缀出浓浓的巴黎风情，无需多余装饰，浪漫风情尽显。